High Electrical Resistant Materials

Scrivener Publishing
100 Cummings Center, Suite 541J
Beverly, MA 01915-6106

Publishers at Scrivener
Martin Scrivener (martin@scrivenerpublishing.com)
Phillip Carmical (pcarmical@scrivenerpublishing.com)

High Electrical Resistant Materials

Ferrochrome Slag Resource Ceramics

Muktikanta Panigrahi
Ratan Indu Ganguly
and
Radha Raman Dash

Scrivener Publishing

WILEY

This edition first published 2024 by John Wiley & Sons, Inc., 111 River Street, Hoboken, NJ 07030, USA and Scrivener Publishing LLC, 100 Cummings Center, Suite 541J, Beverly, MA 01915, USA

For more information about Scrivener publications please visit www.scrivenerpublishing.com.

Wiley Global Headquarters
111 River Street, Hoboken, NJ 07030, USA

For details of our global editorial offices, customer services, and more information about Wiley products visit us at www.wiley.com.

Library of Congress Cataloging-in-Publication Data

ISBN 978-1-394-23125-6

Cover image: Pixabay.Com
Cover design by Russell Richardson

Set in size of 11pt and Minion Pro by Manila Typesetting Company, Makati, Philippines

Printed in the USA

10 9 8 7 6 5 4 3 2 1

Contents

Preface

Since waste materials are currently endangering our environment, ways of utilizing them have become a global challenge. Currently, R&D work is being carried out to utilize these materials for producing value-added products. In the present investigation, an effort has been made to utilize fly ash (FA) and pond ash (PA) (waste materials from thermal power plants) and high-carbon ferrochrome (HCFC) slag (by-product of the ferrochrome industry) for producing a novel material called ceramics. Kaolin/K-feldspar is mixed with PA/HCFC slag to produce ceramics with formation of mullite.

The FA/PA/HCFC slag-based ceramics can replace porcelain-based ceramics, and some permanent ceramic structures can be constructed with such wastes. Thus, it is no wonder that scientists are trying to develop ceramics utilizing waste materials.

The present text will highlight the mechanism involved in the formation of ceramics from FA/PA/HCFC slag. This will perhaps be the first attempt to use FA/PA/HCFC slag for developing ceramics. Properties and structures made with ceramics are found to be comparable with those made with porcelain-based ceramics. Systematic investigations are currently being carried out to develop new ceramic products using these FA/PA/HCFC slag materials. Performances of these materials above ambient temperature are being evaluated, and results indicate the possible replacement of porcelain with these newly invented ceramics.

The Authors
December 2023

1

Fundamentals of Ferrochrome (FeCr) Alloy and Its Slag

Muktikanta Panigrahi[1]*, Ratan Indu Ganguly[2] and Radha Raman Dash[3]

[1]*Department of Materials Science, Maharaja Sriram Chandra Bhanja Deo University, Keonjhar Campus, Odisha, India*

[2]*Department of Metallurgical Engineering, National Institute of Technology, Rourkela, Odisha, India*

[3]*National Metallurgical Laboratory (NML), Jamshedpur, Jharkhand, India*

Abstract

The manufacture of alloy steel consumes chromium in large quantities. Therefore, there is an ever-increasing demand for chromium. Manufacturers use chromium in the alloy steel used for tool steel, high-speed steel, stainless steel, and ball-bearing steel. In addition, nichrome (Ni-Cr) alloy is used as a heating element for electric furnaces.

Chromium is a ferrite stabilizer and a strong carbide former. Addition of chromium improves strength, oxidation resistance, and corrosion resistance properties. Therefore, the global production of chromium has gained importance over the years.

Chromium is not obtained in its free state; therefore, chromium ore is reduced by normal smelting process. Carbothermal reduction of chromium ore enables production of ferrochrome. Usually, an electrical furnace is used for production of chromium, which consumes high electric energy. As a result, the cost of ferrochrome production is enhanced. During the production, chromium is lost through formation of slag. Hence, ferrochrome slag is thought to be a useful raw material for production of high temperature insulation bricks, construction work using geopolymer made from chromium slag, etc. Therefore, R&D scientists are trying to find suitable uses for other applications.

In this chapter, the fundamentals of ferrochrome slag chemistry are discussed. Results obtained by different workers are cited and thermodynamics models used by different groups of researchers are briefly discussed. Physicochemical properties

**Corresponding author*: muktikanta2@gmail.com

Muktikanta Panigrahi, Ratan Indu Ganguly and Radha Raman Dash (eds.) High Electrical Resistant Materials: Ferrochrome Slag Resource Ceramics, (1–62) © 2024 Scrivener Publishing LLC

are also described which may be helpful for future R&D work. Important consideration is also given to environmental issues. Chromium is present in two ionic forms, i.e., trivalent and hexavalent chromium, with the hexavalent form being detrimental to human health. Therefore, this chapter also discusses how unutilized chromium slag pollutes the environment, making it a human health hazard.

Keywords: Chromite ore, ferrochrome production process, ferroalloys, slags, environmental aspects of slag, compositions of slag, thermodynamics of slags, oxidation state of slag, slag properties (i.e., viscosity, electrical resistivity)

1.1 Introduction

Industrialization is a process of socioeconomic development. However, industrial activities are associated with generation of wastes. Production and dumping of wastes pollute our environment as well as water due to the leaching effect when these wastes are submerged in water [1, 2]. Hence, appropriate actions are needed to safeguard atmospheric pollution and subsurface contamination. In recent times, increased construction activities that have arisen due to industrial and population growth have resulted in the scarcity of available conventional construction materials. Hence, nonhazardous wastes are often utilized for construction work [3, 4]. This will reduce waste disposal problems and the costs associated with construction work.

Ferrochrome slag is such a waste material which is obtained from ferrochromium (FeCr) industries [5]. The FeCr alloy is used in the stainless-steel manufacturing process [6]. Chromium in stainless steel improves properties such as corrosion/oxidation resistance, hardness, tensile strength at elevated temperatures, wear and abrasion resistance, etc. [7].

The FeCr alloy is manufactured from its ore by reducing it with coke. The process is carried out at 1,500 °C in an electric arc furnace. Molten slag and metal are slowly tapped in a mold and subsequently cooled in air to enable formation of stable, dense, crystalline phase [8].

The main constituents of FeCr slag are SiO_2, Al_2O_3, FeO, Cr_2O_3, CaO and MgO [9]. Slag products are crushed and ground into sizes ranging between 6.3 mm and 300 mm. Hence, with increased production of ferrochrome, FeCr slag production (HC FeCr, 52% Cr) has increased.

Mineral chromite with chemical composition $FeCr_2O_4$ (ferrous chromic oxide), is a submetallic mineral belonging to the spinel group (with a generic formula of $R^{+2}O.R^{+3}O_4$). It is the only economic mineral mined for chromium production. Because of the high heat stability of chromite, it is also used as a refractory material for high temperature vessels such as furnaces [10]. Two main products are achieved from refining of chromite: ferrochromium and metallic chromium [10]. The smelting operation must be carried out with chromite ore in order to produce the two products mentioned above. One of the major problems encountered during chromite smelting is the issue of energy consumption (i.e., electricity). A large amount of energy is required to smelt chromite to produce ferrochromium or metallic chromium. According to Keesara [11], approx. 4,000 kWh of energy per ton material weight is required for smelting of chromite. This is due to the high melting temperature of chromium. Regarding economic benefits, revenue generated from a ferrochromium plant is dependent on Cr/Fe ratio [12]. The higher the ratio, the higher the revenue earned by the plant. Thus, the economy of the total process depends largely on quality of ore, cost of energy, transportation cost, and others. Hence, one has to consider these factors before venturing into or manufacturing chromium from available ore. During chromite smelting, energy requirement and its cost depend to a large extent on the technology used in smelting the ore. New technologies have been developed to reduce the energy required for smelting chromite and producing ferrochrome. These technologies, together with the conventional chromite smelting technique, are subsequently discussed in this chapter.

1.2 Chromite Smelting Technologies

Production of ferrochrome involves four primary processes: Conventional smelting process, Outokumpu process, DC-arc process, and Premus process. The aim of these processes is to smelt chromite ore in a submerged electric arc furnace. These technologies are discussed below.

1.2.1 Conventional Smelting Process

Traditional chromite smelting technology involves charging chromite ore into a submerged electric arc furnace (Figure 1.1). Coke and coal are used as reductants. Quartzite is used as a fluxing material which removes undesirable oxides in the form of slags. Metal/ferrochrome and slag produced are tapped from the furnace for further processing. According to Naiker [13], the primary advantages of the conventional smelting process are the low capital investment incurred and the flexibility offered in choosing raw materials in the production process. However, in terms of energy requirement, the process is not efficient as it is an energy-intensive process, requiring up to 4,000 kWh per ton of material produced. The conventional

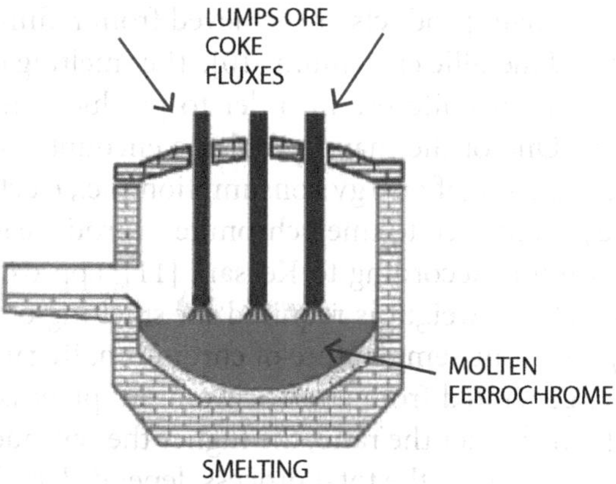

Figure 1.1 Schematic diagram of smelting using submerged electric arc furnace (EAF) [14].

smelting process requires about 1.0 indexed energy cost per ton of ferrochrome alloy produced.

1.2.2 Outokumpu Process

The Outokumpu process involves grinding and pelletizing of ore fines, followed by sintering of green pellets and preheating before smelting [15, 16]. According to Goel [16], the ore and coke fines are normally wet-ground to about 35 percent under 37 microns (400 meshes) and are then pelletized to approximately +15 mm size. Figure 1.2 shows the flow sheet of the Outokumpu process. As seen in Figure 1.2, the preheating operation is done mainly in a rotary kiln and the energy required for sintering and preheating the pellets comes from CO gas generated in a submerged arc furnace. In the Outokumpu process, chromite ore is partly reduced in the rotary kiln during preheating, thereby reducing the amount of electric energy required for final smelting of the ore to ferrochrome.

1.2.3 DC-Arc Process

DC arc furnaces use a single solid carbon electrode as the cathode and produce a DC arc to an anode in the bottom of the furnace (Figure 1.3). The arc is normally an open or semi-submerged one [18]. Raw materials can be charged either directly into the furnace or by using a hollow electrode. The DC arc route is designed mainly to overcome the problem of ore fines encountered in the conventional chromite smelting process. Therefore, in

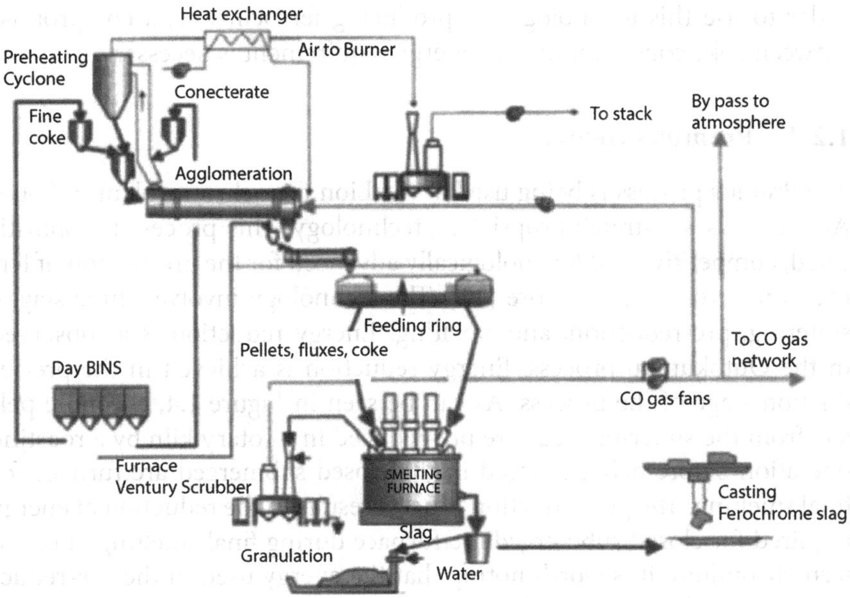

Figure 1.2 Outokumpu process [17].

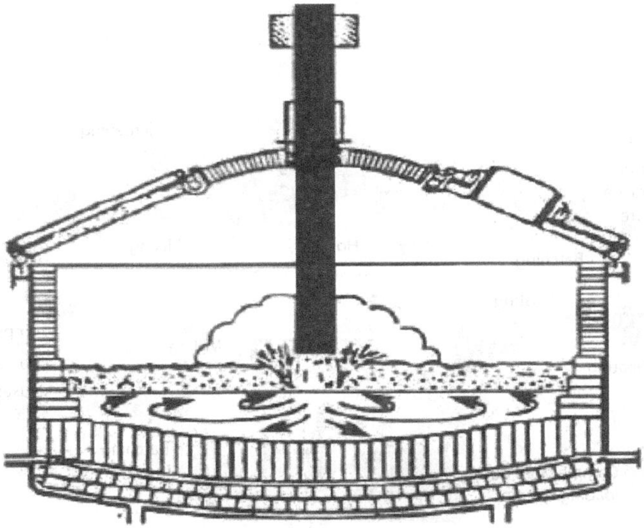

Figure 1.3 DC-arc process [19].

order to use this technology for producing ferrochrome, a compromise between coke consumption and energy requirement is necessary.

1.2.4 Premus Process

The Premus process is being used at the Lion ferrochrome plant in South Africa. It uses Xstrata's proprietary technology. This process is sophisticated, competitive and technologically advanced for the production of ferrochrome from chromite ore [17]. The technology involves three stages: sintering, pre-reduction, and smelting. Energy reduction is as observed in the Outokumpu process. Energy reduction is achieved in the pre-reduction stage of the process. As can be seen in Figure 1.4, chromite pellets from the sintering stage are pre-reduced in a rotary kiln by a roasting operation before being charged in the closed submerged arc furnace for final smelting. The pre-reduction process results in the reduction of energy required for closed submerged arc furnace during final smelting of ore to ferrochromium. It is worth noting that the energy used in the pre-reduction stage is obtained from hot gas generated in the closed submerged arc furnace and also from coal pulverization, and thus is not paid for. With this technology, Lion ferrochromium is now in the forefront of chromite smelting in South Africa.

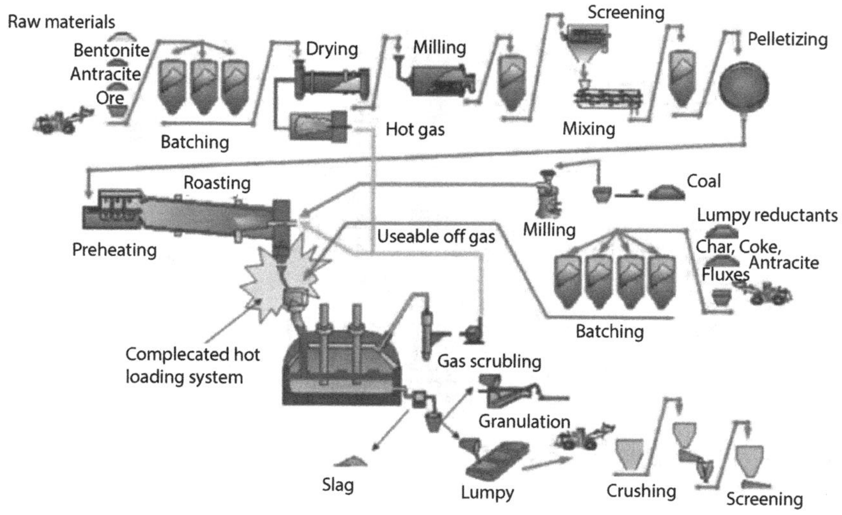

Figure 1.4 Schematic illustration of Premus process [13].

Premus technology is the least expensive and most energy-efficient ferrochrome technology. The technology also uses low-cost reductant material (anthracite), and thus coke consumption is low in Premus technology.

Another important feature of this technology, which is also a part of the DC-arc process and Outokumpu process, is that it is designed for smelting fine chromite ore. The conventional smelting process is designed to handle only lumpy ores. This is ideal for production of ferrochromium as the fine ores tend to form a sintered layer at the top of the charge. Thus, using fine ores sometimes causes a technical problem, such as preventing gas from escaping, thus posing operating difficulties [12]. Due to this problem, fine chromite ores have low economic value. However, with the development of Premus technology, DC-arc process, and Outokumpu process, fine chromite ores can now be smelted. These technologies have been designed to convert the fine ores into pellets during the pre-reduction stage. These technologies can therefore be used to generate more revenue in localities having a large amount of fine chromite ore.

1.3 Ferrochrome Production

Ferrochrome is produced by carbothermic reduction of iron magnesium chromium oxide at high temperature. Coke and coal are used as reducing agents. The reduction process is carried out in an electric arc furnace. The arc generates temperature around 2800 °C. For this purpose, the electricity consumed is high and consequently cost increases [20].

1.4 Chemical and Physical Properties of Ferrochrome

Ferrochrome or ferrochromium (FeCr) is a type of ferroalloy. Usually, such alloys contain chromium (50 to 70% by weight) and iron [22, 23]. Physical and chemical properties of ferrochrome are listed in Table 1.1.

Globally, large quantities of domestic chromite resources are available in South Africa, Kazakhstan and India. Stainless steel manufacturers are the largest consumer of chromium and ferrochrome. FeCr production (High Carbon FeCr, 52% Cr) has now reached the level of 5 M ton/year [24].

Chromite ore is ferrous chromic oxide ($FeCr_2O_4$). It belongs to spinel group with a generic formula of $R^{+2}O.R^{+3}O_4$. FeCr alloy is produced from chromite ore. Because of its high heat stability, it is used as a refractory liner of an arc furnace vessel.

Table 1.1 Physical and chemical properties of ferrochrome [21].

S.no.	Physical properties	Physical property value
1	Atomic Mass	52.0
2	Density	7.15 g/cm^3
3	Melting Point	1900 °C
4	Boiling Point	2642 °C
5	Appearance	Ferrochrome is available in a variety of forms, including small crystals, lumps and granules as well as in powder form.
6	Color	Color varies from dark gray to light gray.
7	Odor	Odorless; can be dangerous when inhaled.
8	Solubility	The alloy is not soluble in water.
9	Combustibility	The dust particles of this chemical alloy are combustible.

Ferrochrome is collected from the furnace at regular intervals. Usually, slag and molten metal are tapped where molten metal is separated in molten condition and is chilled in a metal mold. The molten ferrochrome is then solidified in large castings and is processed further.

On the basis of carbon content, ferrochrome alloy is classified into two categories: low carbon ferrochrome alloy and high carbon ferrochrome alloy.

1.4.1 Characteristics of Ferrochrome

- It is a steel-gray and hard metal.
- Chromite ore minerals belong to the spinel group.
- Cr is a carbide-forming element and forms $Cr_4C_3/Cr_{23}C_6$ in stainless steel.
- Cr is a ferrite stabilizer and makes the alloy stable at high temperature.
- Cr in stainless steel improves corrosion-oxidation resistance.
- Cr improves stress corrosion property in stainless steel.
- Metallic chrome products or super alloys or other special melting metallic products are produced using chrome alloy.

1.5 Ferrochrome Uses

Most of the ferrochrome produced worldwide is used in manufacturing of stainless steel. Chromium content stainless steel varies between 8 to 29% FeCr. Chromium content stainless steel provides resistance to corrosion. Ferrochrome is indispensable in manufacturing of ball-bearing steels, tool steels and other alloy steels [25]. Apart from making stainless steel, low-carbon ferrochrome is also used in the manufacturing of acid-resistant steels [26].

Ferrochromium alloy picks up carbon during carbothermal reduction. Since Cr and Fe have affinity for carbon, they pick up carbon. The amount of carbon in alloy depends on operating conditions, i.e., oxygen potential and temperature. Two types of ferrochrome alloy are produced [27, 28];

1. Low-carbon ferrochrome alloy
2. High-carbon ferrochrome alloy.

1.5.1 Low-Carbon Ferrochrome Alloy

An optical photograph of low-carbon ferrochrome is shown in Figure 1.5. Low-carbon ferrochrome is a ferrochrome having carbon content below 0.25. This is used as raw materials for producing silicon chromium alloy,

Figure 1.5 Optical photograph of low-carbon ferrochrome [29].

chromite and lime. Low-carbon ferrochrome chromium (LC FeCr) is used in austenitic stainless steel and superalloy. Due to high quality, it is a reliable and economical alternative to metallic alloy chromium for superalloy production [29].

1.5.1.1 Low-Carbon Ferrochrome - Chemical Analysis

Ferrochrome is an alloy of chromium and iron containing 50–70% chromium by weight. Addition of chromium improves the strength and yield point of steel and reduces elongation insignificantly. The presence of chromium in carbon steels improves hardness and wear resistance. The chemical composition of low-carbon ferrochrome is indicated in Table 1.2.

1.5.2 Problems of Carbon in FeCr

Ferrochrome production is done by chromite reduction with carbon (coke) in a submerged arc furnace. Ferrochrome product is basically carbon-saturated iron-chromium alloy with 6–8 wt% carbon. Carbon forms solid chromium carbide and complex iron-chromium carbides. The higher percentage of carbon in ferrochrome creates several problems.

Low-carbon alloy for making austenitic stainless steel, carbon contains less than 0.1% (0.02–0.05%). Excess carbon causes "sensitization" problems at about 810 K. At sensitization temperature, excess carbon combines with chromium to form chromium carbide ($Cr_{23}C_6$). At this temperature, carbide is insoluble and precipitates at the grain boundaries. At grain

Table 1.2 Chemical composition of low-carbon ferrochrome (LCFeCr) [30].

LCFeCr	Low-carbon ferrochrome (60%)	Low-carbon ferrochrome (70%)
Cr	65% Min	70% Min
C	0.03, 0.06, 0.10, 0.15	0.03, 0.05, 0.06, 0.08, 0.10, 0.15
Si	1% Max	1% Max
S	0.03% Max	0.03% Max
P	0.035% Max	0.035% Max
Size	10–50 mm – Min 90% or as required by end user	

boundaries, chromium has inadequate corrosion resistance. Thus, carbon is indirectly responsible for corrosion, and hence weakens the material. A similar problem is observed in welded structures [28, 31].

In austenitic stainless steels, the chromium to carbon ratio is more than 150, whereas in high-carbon ferrochrome steel it is about 10. High-carbon ferrochrome stainless steel inadvertently increases carbon content. Hence, it is essential to remove carbon from the parent source (i.e., high-carbon ferrochrome). To improve the chemical and mechanical properties of steels, minimizing the concentration of the detrimental and undesirable element (i.e., carbon) from steel is necessary.

Chromium metal is reactive towards carbon by simple oxidation to form chromium oxide. Such practice cannot be used for removing carbon. This is a major problem in the case of most ferroalloys [32].

Conventional decarburization techniques are used to remove carbon from FeCr. There are many disadvantages to these techniques such as high refractory consumption, poor metal recovery due to losses of chromium in slag, etc. However, nonconventional decarburization techniques are not used. Conventional decarburization methods are described below.

1.5.2.1 Refining of Ferrochrome by Chromium Ore

Elyutin [32] assumed that liquid high-carbon ferrochrome is a solution of chromium and carbon which exists in the form of carbide (i.e., Cr_7C_3). This process is used to determine the thermodynamic characteristic. Decarburization reaction is written as (Eq. 1.1):

$$\begin{aligned} &\langle Cr_2O_3\rangle + \langle FeCr_2O_4\rangle \rightarrow \langle Cr\rangle + \langle Fe\rangle + CO \\ &\Delta G^o = 612 - 0.323T \quad kJ \end{aligned} \tag{1.1}$$

Difficulties encountered in the decarburization process are due to the high melting point and viscosity of the high-chromium refining slag. This prevents the growth of chromic oxide in the slag. Hence, there is a limited possibility of increasing the oxidizing capacity of slag. Decarburization process is conducted at a temperature of 2175 K; 2% of the carbon in the alloy is decarburized. Decarburization of low carbon grade material is feasible at a temperature higher than 2375 K. Thus, production of low-carbon ferrochrome by this process is very difficult.

1.5.2.2 Refining of Ferrochrome by Blowing Oxygen

In this process, carbon is removed by blowing O_2 through molten ferrochrome in a converter. It is a more expensive process. Elyutin [32] brought carbon below 1% using air as an oxidizer. This method did not give satisfactory results because of insufficient bath temperature (i.e., 1935–1985 K).

1.5.2.3 Refining of Ferrochrome in Presence of Silicon

The process consists of refining ferrochrome by oxygen in the presence of silica under high vacuum. Since the reaction is carried out, gaseous product is formed. The disadvantages of the process are contamination of alloy with unreacted silica, high silicon content of product and lengthy production time.

1.5.2.4 Silicothermic Process (Production of Low-Carbon Ferrochrome)

Increasing silicon content in an alloy decreases solubility of carbon. Low-carbon ferrochrome is prepared by reduction of chromite using silicon as reducing agent and silico-chrome is formed. Smelting is done in a shaft type of electric furnace lined with magnesia [33]. Size of chromite ore should be less than 50 mm and size of ferrochrome should be less than 10 to 15 mm. Also, chromite ore contains less phosphorus and sulfur. Ferrosilicon chrome is free from slag inclusions. This is because slag contains entrapped carbon in silicon carbide particles. Use of higher voltage minimizes carbon pick up from the electrodes. The main advantage of the process is that it can produce an alloy with carbon less than 0.03%.

1.5.2.5 Production of Carbon-Free Ferrochrome (Aluminothermic Process)

In this technique, carbon-free ferrochrome is produced on a small scale. The reduction reaction (Eq. 1.2) proceeds as follows;

$$\frac{2}{3}\langle Cr_2O_3\rangle + \frac{1}{2}\langle Al\rangle = \frac{4}{3}\langle Cr\rangle + \frac{2}{3}\langle Al_2O_3\rangle \qquad (1.2)$$
$$\Delta G^o = -309 + 0.004\,T \quad kJ$$

The amount of heat produced per unit mass of reactants is 2649 kJ/kg, which is sufficiently higher [34]. Aluminothermic reduction will be easier in comparison to the silicothermic process.

1.5.3 Non-Conventional Techniques

Non-conventional techniques (AOD and VOD) have been adopted for the stainless-steel manufacturing process since several difficulties are encountered by conventional decarburization methods.

1.5.3.1 Decarburization of Solid Ferrochrome

0.01 to 0.03 % carbon content ferrochrome is obtained from high-carbon ferrochrome (6–8% carbon) by vacuum heat treatment process with an oxidant (e.g., iron oxide). A distinctive composition of ferrochrome of 60–70% Cr, 0.01–0.03% C is obtained by this process.

1.5.3.2 Decarburization Using Oxidizing Gas Mixture

(a) Water Vapor
Oxidation of carbon from ferrochrome takes place according to the reaction shown in Eq. 1.3. Shimizu [35] carried out the experiment using water and calcium compound to form dense ferrochrome. Actual mechanism of decarburization is still not clear. This is because of the endothermic nature of the reaction (Eq. 1.3) and requirement of high vacuum.

$$\begin{aligned} &\langle Cr_{23}C_6\rangle + 6(H_2O) = 23\langle Cr\rangle + 6H_2 + 6(CO) \\ &\Delta G^o = 1890 - 0.296\,T \quad kJ \end{aligned} \tag{1.3}$$

(b) Carbon Dioxide
In this method, decarburization is carried out by subjecting ferrochrome pallet to carbon dioxide flow. Decarburization proceeds by the following reaction (Eq. 1.4):

$$\begin{aligned} &\langle Cr_{23}C_6\rangle + 6(CO_2) = 23\langle Cr\rangle + 6H_2 + 12(CO). \\ &\Delta G^o = 1436 - 1.012\,T \quad kJ \end{aligned} \tag{1.4}$$

It is basically a solution reaction, which is endothermic. Thermodynamically high temperature and low partial pressure of CO can help make it feasible [36].

1.5.3.3 Production of Low-Carbon Ferrochrome from Chromite Ore

Other than decarburization of high-carbon ferrochrome, low-carbon ferrochrome material is processed by reduction of chromite ore through solid carbon (i.e., graphite) or H_2 to form metallic chromium or an alloy of iron and chromium in solid state.

Chu [37] has studied reduction experiments with H_2 of chromite ore. Rankin [38] has reduced chromite ore by CO. He analyzed the diffraction pattern of the product at various stages and at different temperatures.

Sundermurti [39] carried out chromite reduction in Stanton Red croft thermo-gravimetric balance under inert environment (i.e., argon gas). Also, he studied the kinetics of the process. Hiroshi [40] reported the fractional reduction of iron and chromium oxide. However, he has not noticed carbide formation.

Synthetic chromite reduction was studied by Sheshadri [41]. Various mechanisms were explored to derive a rate-controlling step. However, he found that these processes are kinetically slow. Additionally, he observed that the separation of metal beads is difficult from unreduced oxide and gangue material.

1.5.3.4 Khalafalla Method

Pahlman [42] used a direct method to produce low-carbon ferrochrome. Khalafalla [43] has suggested a direct method to prepare low-carbon ferrochrome alloys containing carbon less than 2%. In his process, chromite is reduced through electrothermal reduction route. The reduced product contains less than 2% carbon. The process of reduction of chromium pellet is carried out using carbonaceous reductant mixture at a pressure of 0.1 to 1 torr and temperature of 1500 to 1600 K. Reduction of chromite to metallic product is found to proceed via formation of intermediate carbides (i.e., $Fe_3C + Cr_3C_2$). Soaking the charge for about 20 minutes has given button of low-carbon ferrochrome alloy [43].

1.5.3.5 Other Methods

In low-carbon ferrochrome, efforts are being made for decarburization of carbon in pre-reduced chromium ore by J.E. Stress [40]. Pre-reduced chromium ore reduces power consumption and increases ferrochrome.

K. Suzuki has studied decarburization of liquid Fe-Cr-C alloys by rotating Cr_2O_3 cylinder in the melt [44].

Low-carbon ferrochrome is being produced worldwide by various processes, namely [45]:

1. Perrin process
2. Duplex process
3. Simplex process
4. Triplex process
5. Fusion process.

1.5.4 High-Carbon Ferrochrome Alloy

The metallic Cr content of high-carbon ferrochrome alloy is 60–65% with varying amounts of Fe and C. Therefore, quality control of raw materials ensures maximum output and uniform quality of the product. The main components of slag are SiO_2, MgO, and Al_2O_3. Slag also includes oxides of Cr and Fe and calcium [9, 46]. Common phases in slag are glass, spinels (Al_2O_3–MgO) and forsterite (MgO–SiO_2) and a small amount of CaO [47].

Silicon content in conventional high-carbon ferrochrome is normally below 2%. Thermodynamic equilibrium allows flexibility of silicon content (2% and 5%) in the high-carbon ferrochrome alloy. Silicon content of alloy depends on chosen production technology and operating conditions. An optical photograph of high-carbon ferrochrome is shown in Figure 1.6.

Figure 1.6 Optical photograph of high-carbon ferrochrome [48].

Table 1.3 Typical chemical composition of high-carbon ferrochrome [49].

HCFeCr	High-carbon ferrochrome
Cr	60% Min
C	8% Max
Si	3% Max
S	0 . 04% Max
P	0 . 04% Max
Cu	0 . 5% Max
Size	10–150 mm; Min 90% or as required by end user

High-carbon ferrochrome is an intermediate product and is used as a feed material for production of stainless steel. Characteristic features of high-carbon ferrochrome include crushed lumpy product with a silvery metallic appearance and irregular, flaky granulated product with a greenish appearance. The chemical composition of high-carbon ferrochrome is given in Table 1.3.

An optimal slag composition is achieved using a modern process. Factors like temperature and composition of slag, viscosity and electrical conductivity are prime criteria. Major compositions of slag are SiO_2, MgO and Al_2O_3.

Also, slag contains Cr oxide, Fe oxides and Ca oxide. Phases present in slag are glass, spinels and forsterite. Slag chemistry is important to get an efficient amount of ferrochrome.

1.6 Waste Generation During Ferrochrome Production

Chromite ore contains undesirable oxides. These oxides, such as chromium, magnesium, iron, and calcium oxides, are called gaunge materials. During production of FeCr alloy, gangue materials are removed by addition of fluxing agent (i.e., quartzite). Thus, slag generated from the smelting furnace contains MgO, CaO, Cr_2O_3, SiO_2, Al_2O_3, etc.

Typical reduction reactions in the ferrochrome furnace can be seen in Figure 1.7. Other smelted products are CO gas and FeCr slag. Carbon monoxide (CO) gas is obtained from reduction reactions. It is generated at 700 Nm^3/t FeCr. In high temperature, charge materials start to smelt. Materials are not dissolved into metallic ferrochrome phases and primarily

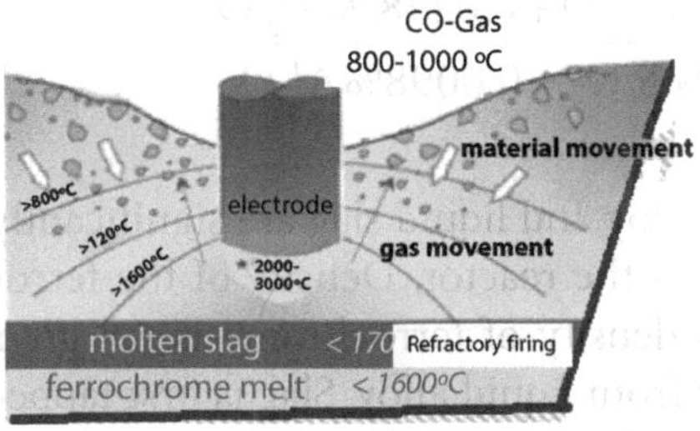

1.800°C

$Fe_2O_3 + CO \rightarrow 2\,FeO + CO_2$

2.1200°C

$FeO + C \rightarrow Fe + CO$

3.1400°C

$Cr_2O_3 + 3C \rightarrow 2Cr + 3CO$

4.1600°C

$SiO_2 + C \rightarrow SiO + CO$

Figure 1.7 Typical reduction process in the ferrochrome furnace [50].

silicate phases are formed. Silicate phases are generated in the form of ferrochrome slag. Ferrochrome slag production is 1.2 t / t FeCr. It is mainly in granulated form [50].

1.6.1 Ferrochrome Slag

The FeCr Industries generate a huge amount of wastes. These wastes are dumped, which pollute our environment. Due to the leaching effect, underground water is polluted. Therefore, appropriate precautions are to be taken in order to save our contaminated environment. Currently, there is increased construction activity due to industrial and population growth. These activities have resulted in scarcity of conventional construction materials. Hence, it has led to an increase in construction costs.

Some nonhazardous wastes are suitable for use as construction material. This will not only solve the waste disposal problem, but also reduce the cost of construction.

Major components of FeCr slag are SiO_2, Al_2O_3, FeO, Cr_2O_3, CaO and MgO [50]. A major part of slag is granulated and sizes vary between 6.3 mm to 300 mm.

Temperature of slag and ferrochrome are tapped at 1700 °C and 1600 °C, respectively. Smelting temperature of slag is higher than temperature in metal alloy. This is because metal alloy is heated up in slag liquid phase. Optimum smelting temperature is practically noted between 1680–1720 °C. The slag composition is checked in every melting process. A proportion of quartzite in the charge mix controls the slag composition, which is tested every day by sampling and analysis. Ferrochrome slag composition is SiO_2 (30 %), Al_2O_3 (26 %), MgO (23 %) and CaO (2 %). Chrome and iron contents in the slag are 8% and 4%. Ferrochrome slag is acidic in nature. Its basicity is 0.8. Basicity is calculated by the following formula:

$$B_3 = \frac{0.0178.\%CaO + 0.0248.\%MgO}{0.0166\%SiO_2 + 0.0098\%Al_2O_3}.$$

Since the densities of metal alloy and liquid slag are significantly different, slag is easily separated from the reactor. Density of the ferrochrome slag is 2.5–2.8 g/cm^3, whereas density of ferrochrome is 6.8 g/cm^3. This makes separation of slag easier from liquid alloy. Slag is first tapped in the upper part of the reactor. Liquid alloy is tapped in the lower part of the rector. Hence, separation occurs easily. Viscosity is also important during separation of slag and metal alloy. Viscosity of ferrochrome slag will increase when the SiO_2 and Al_2O_3 content rise. Total chemical composition of slag defines the order of crystallization during the cooling process. The cooling rate has an influence on degree of crystallization. It is crucially important for the operation of the smelting furnace [50].

1.6.2 Ferrochrome Slag Products

Slag produced from a ferrochrome smelter will have different characteristics depending upon composition and mode of cooling of molten slag. Slag is tapped either in ladles or in launders. If it is tapped in ladles, then the slag produced becomes granulated. The overflow from the ladles that flows along the slag launder to a pond is depicted in Figure 1.8.

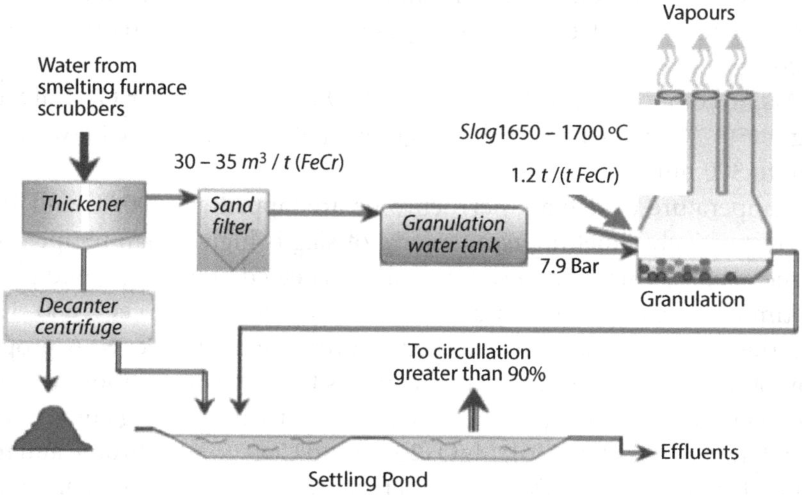

Figure 1.8 Slag granulation process [50].

Ferroalloy separation from lumpy slag is based on gravity and magnetic separation process. The first stage of slag processing is crushing and screening of feed material to fractions of 0-4 and 4-22 mm. Coarser fractions are handled by heavy media separation (DWP method). A magnetic separator and spiral washing are used for fine-grained material. Recovered slag of 0-4 and 4-22 mm are saleable products. The slag products from the process are aggregates of 0-4, 4-11 and 11-22 mm. Crystallization portion of lumpy slag is higher due to slower air cooling than that of granulated slag. Structure of slag is partly crystalline and partly glassy. Significant phases are amorphous glass, Fe-Mg-Cr-Al-spinels, forsterite, Mg-Al-silicate and metal alloy [50].

1.6.2.1 Petrographic Analysis of Low-Carbon Ferrochrome Slag

Complex processing of industrial wastes is an integral part of development for ferroalloy industries. An example of such an approach is extraction of alumina from ferroalloy slags having high content of aluminum oxide. Fe-Cr-Si (used as a reductant) is used for low-carbon ferrochrome alloy. Fe-Cr-Si may be replaced by less expensive ferrosilicoaluminum (Fe-Si-Al). In such a process, slag is formed having a composition near to the composition of helenite phase ($2CaO{\cdot}Al_2O_3{\cdot}SiO_2$).

Helenite slag containing more than 25% of Al_2O_3 is a resourceful raw material for production of alumina (construction materials). Petrographic analysis has revealed the presence of different forms of helenite phases. Isolated impregnations of melilite, larnite and vitreous phase are distinctly separable, which is proof of their extraction [51].

Phase compositions of slag used in the studies were determined to be Cao-MgO-SiO_2, 2CaO-MgO-2SiO_2, 2CaO-Al_2O_3-SiO_2, MgO-Al_2O_3, MgO-Cr_2O_3, FeO-Cr_2O_3 and 3CaO-MgO-2SiO_2 [52, 53]. Petrographic analyses are carried out for polished section of samples using Neophot 21 microscope. Slags are related to microcrystalline types which include helenite and anorthite phases. Sample phases are mostly represented by cryptocrystalline helenite $2CaO{\cdot}Al_2O_3{\cdot}SiO_2$.

1.6.2.2 Description of Crystalline Phases in Ferrochrome Slag

Different types of phases are present in ferrochrome slag. The actual nature and composition of slag depend on the chemical composition of ore, fluxing agents, and reduction parameters such as oxygen partial pressure, reduction temperature, and residence time in high temperature zone [54]. The melting and crystallization process in ferrochromium slag is very important for understanding the developed crystal phases. Formation of phases affects properties such as strength, refractoriness, bulk density, and chromium stability in slag. Phases present are CaO-A_2O_3-MgO-SiO_2 and MgO-MgO.Al_2O_3-2MgO.SiO_2-2CaO.SiO_2 systems.

During the cooling period, crystals of MgO, MgO-spinel, MgO-spinel-forsterite crystallize from the molten slag [55, 56]. Hayhurst [55] has studied spinel-forsterite-anorthite-coerdite system to understand the crystallization process of high-carbon ferrochromium slag during cooling. During cooling of high-carbon ferrochromium slag, spinel and forsterite phases crystallize from solution in liquid phase. Formation of glassy and enstatite phases are due to incomplete peritectic reactions during rapid cooling. In fully crystallized ferrochromium slag, spinel phases are dissolved in silicate solution at 1340 °C. This temperature is reduced to 1280 °C due to the presence of coerdite phase. Presence of Cr_2O_3 in spinel phase increases smelter temperature. Thus, it is concluded that major phases in ferrochrome slag are chromite spinels, forsterite, enstatite, anorthite; and minor phases are spinel and chromium carbide [55].

Particles of unreduced chromium (PAC) are surrounded by angular spinel of low reflectivity. It contains globules of metallic ferrochromium. Lath-shaped crystals of enstatite are dispersed in glassy matrix. Calcium oxide to silicon dioxide ratio (CaO/SiO_2) and magnesium content in ferrochrome slag are two major factors to control the leaching behavior of chromium. The critical Cr-bearing mineralogical phases formed under certain criteria are as follows:

(a) If CaO/SiO_2 ratio is less than 2, then Cr preferably exists in slag as magnesiochromite spinel ($MgCr_2O_4$) phase.
(b) If CaO/SiO_2 ratio is greater than 2, then Cr preferably exists in slag as calcium chromite. Calcium chromite phase is susceptible to leaching by acid and may be oxidized to calcium chromate if exposed to the environment for a long period of time. However, magnesiochromite spinel phase is stable against oxidation and is resistant to dissolution in aqueous media [57].

Mutual interference between phases causes interlocking of crystalline phases. Thus, solidified ferrochromium slag is very hard and possesses high refractoriness with enhanced mechanical strength at high temperature. Residual chromium present in ferrochromium slag remains stable and gets immobilized due to its locking with highly stable crystal phases like chromite/magnesium aluminium chromite/magnesiochromite. Therefore, chromium release from slag matrix under normal environmental conditions is inhibited [57].

1.6.2.3 Slag Properties

Slag properties are important since they play a vital role in process control and product quality. Also, they are needed to develop reliable mathematical models of the process and to optimize the efficiency of the process and quality control. Slag compositions have high sulfur capacity (S-capacity), which means maximum amount of transfer of sulfur from metal to slag. This is usually achieved by high-basicity flux. Increased basicity due to addition of CaO will ensure effective removal of sulfur from ore to slag. However, higher basicity will increase the liquidus temperature (T_{liq}) of slag.

Product quality will occur by adoption of continuous casting (CC) process. Longitudinal cracks occur on the surface of steel and have to be scrubbed off. This process will reduce yield and increase expenditure due to energy consumption for scrubbing. In order to minimize longitudinal cracking on the sample, it is essential to optimize variables such as flux viscosity (η) and break temperature (Tbr) of mold fluxes used in the CC process.

There are several types of mathematical models, namely thermodynamic models (to establish viability of process and optimum conditions), kinetic models (to establish productivity rate), heat and fluid flow models (to improve process design and control), and models of the environmental and financial impact of the process. Reliable property values are principally needed as input data for heat and fluid flow models.

1.6.3 Reactivity of Slag with Container (Metal and Atmosphere)

With an increase in temperatures, reactivity increases; therefore, the possibility of a reaction occurring between two systems accelerates exponentially at very high temperatures. Slags react with container and other contact materials (i.e., metal and atmosphere). The extent of reactivity varies between viscously aggressive (e.g., coal slags and ash) to benign (glasses) depending upon its composition.

Some slags are unstable at high temperatures and reactivity between two or more components in the slag react, thereby continuously changing the composition. The following Eq. 1.5 is an example of ESR slags where fluorides and oxides react:

$$3CaF_2 + Al_2O_3 \rightarrow 2AlF_3\,(g) + 3CaO \tag{1.5}$$

1.6.3.1 Slag/Container Reactions

Some slags are aggressive and reactive towards the refractory walls where there process. In such cases, suitable refractory liners are to be identified where slag are formed. A major problem in performing measurements on slags involves the selection of a crucible material which is inert to the slag. The general rule in selecting crucibles is: Use a metal to contain an oxide melt and use a refractory oxide to contain liquid metals. Commonly used materials are either

(i) Pt metals and alloys (e.g., Pt, Ir), or
(ii) Refractory metals (Mo, Ta, and W).

Pt-alloys are not suitable for reducing conditions, and Mo, etc., are not suitable for oxidizing conditions. FeO-containing slags are notoriously aggressive since FeO is easily reduced by the container. Use of Fe crucibles is limited to a temperature range of less than 1400 °C. Graphite is also widely used but slag oxides react with graphite (Eq. 1.6).

$$\text{FeO slag} + \text{C} \rightarrow \text{CO(g)} + \text{Fe} \tag{1.6}$$

Containerless methods have been used to avoid slag/container reactions [58].

1.6.3.2 Slag/Metal Reactions

Slag/metal reactions are very important in the production of clean metals (e.g., de-S and de-P) but they can sometime produce unwanted reactions in continuous casting (CC) (Eq. 1.7).

$$SiO_2 \text{ slag} + Al \rightarrow Si + Al_2O_3 \text{ slag} \tag{1.7}$$

For steels containing > 1% Al, this reaction makes steels virtually uncastable (by CC). Slag/metal reactions involve ions.

1.6.3.3 Slag/Atmosphere Reactions

Slag is used in many processes to shield metal from oxidation. Oxygen can have a significant effect on transition-metal and multivalent oxides (e.g., Fe^{2+}/ Fe^{3+}, Cr^{2+}/ Cr^{3+}, Ti- Nb^{-} oxides). In the case of Fe^{2+}/ Fe^{3+}, the Fe_2O_3 concentration has a significant effect on properties (e.g., viscosity, electrical resistivity). Thus, it is important that the partial pressure of oxygen

(pO_2) in an experiment be selected to reflect conditions of the industrial process being studied. Nitrogen tends to be inert in most slags, but those containing Ti and Nb will react with N_2 to form TiN and NbN inclusions, which can have significant effects on the process.

Reactions of slags with moisture can have important effects. Water (H_2O) remains in slag in two forms:

(i) Chemically-bonded, and
(ii) Adsorbed form.

Performance of slags depends on the extent of water present within it. However, the effects of moisture are more severe when dealing with slags containing fluorides, or chlorides such as ESR fluxes, CC fluxes and Al fluxes. The following reaction (Eq. 1.8) exemplifies the above statement:

$$CaF_2 \text{ slag} + H_2O_g \rightarrow CaO \text{ slag} + 2HF_g \qquad (1.8)$$

The HF_g emitted can cause plant corrosion and pose a threat to the health of plant personnel.

1.6.4 Corrosion and Erosion of Refractories

Corrosion is the deterioration of materials under an aggressive environment. Erosion is a mechanical abrasion of material. This process reduces the life of the refractory lining. Therefore, the refractory lining is chosen from among those materials which have high chemical stability and melting point (MPt). Corrosion is the chemical attack of refractories. Erosion involves both the chemical attack (erosion) and the mechanical abrasion of the refractory (Figure 1.9). Refractories (e.g., Al_2O_3, CaO, MgO and ZrO_2) used for furnace linings are usually oxides with great chemical stability and have high melting points. These properties arise from the great chemical affinity of Al, Ca, Mg and Zr for oxygen (the bond strength Al-O, etc., is very high); these elements lie at the bottom of the Ellingham diagram and it is very difficult for metals to break the Al-O bond and form a metal-oxide bond. However, most slags will dissolve a certain amount of the refractory oxide, e.g., CC fluxes dissolve up to 40% Al_2O_3, up to 10% TiO_2 and only 2% ZrO_2 and 0.5% TiN.

Erosion frequently arises from turbulent flows generated in the reactor, e.g., the basic oxygen steelmaking (BOS) process (Figure 1.9) or from mechanical erosion, e.g., the feeding of scrap to the converter. Enhanced erosion tends to occur in furnace linings at the metal/slag and slag/atmosphere interfaces. This is known as "slag line erosion."

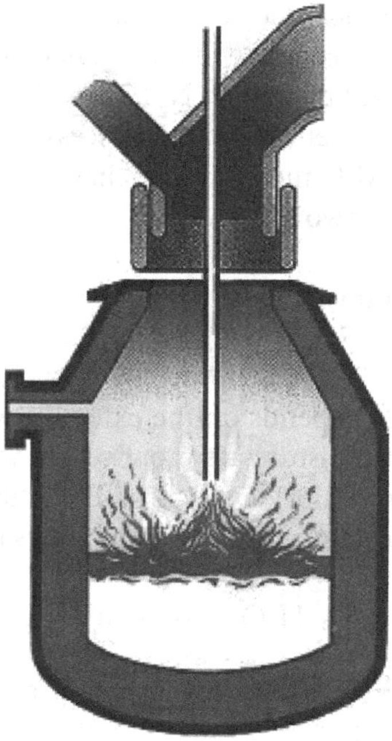

Figure 1.9 Schematic drawing of the BOS process and the turbulence generated, which causes refractory erosion [59].

1.6.4.1 Mechanism of Slag Line Erosion

Refractories used in industrial furnaces frequently exhibit an area of wear (deep groove) in the region between slag/metal and slag/gas interfaces. Figure 1.10 shows the case of a slag line attack at the slag/atmosphere interface for an SiO_2 refractory. The same mechanism is applied for other refractories. When slag first comes in contact with silica, it forms a meniscus. Some SiO_2 dissolves in the meniscus and this alters surface tension (e.g., $\gamma SiO_2 > \gamma PbO$ but $\gamma SiO_2 < \gamma FeO$); thus, for the PbO-$SiO_2$ system, surface tension is greater at the tip than the base of the meniscus. Marangoni flow (surface-tension-driven flow) causes a flow from low to high surface tension so a flow is established which results in a vortex formation. This vortex flow causes gradual erosion of the refractory of the region [60]. The same mechanism applies to FeO-SiO_2 slag but the direction of slag flow will be the reverse of that for PbO-SiO_2.

In the case of PbO-SiO_2, slag film at the meniscus tip is higher than that at the base and the upward meniscus, and the Marangoni flow causes the slag to cover new refractory. This dissolves into a thin slag layer and creates higher surface tension (γ). So, slag is pulled further up to new

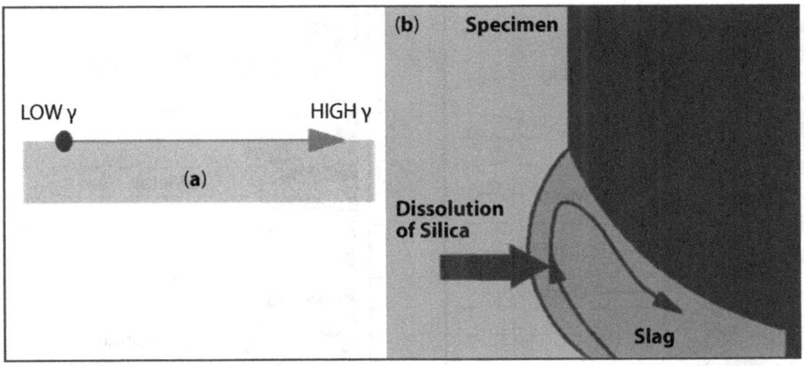

Figure 1.10 Schematic drawings showing (a) direction of Marangoni flow and (b) mechanism for slag line attack in PbO-SiO_2 slag [60].

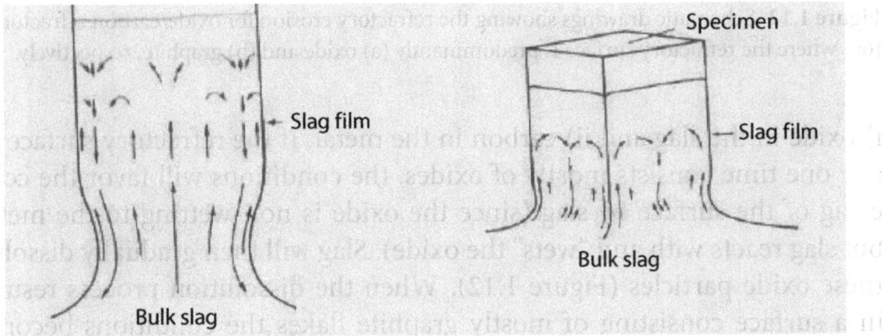

Figure 1.11 Schematic diagrams showing refractory erosion and creeping for PbO-SiO_2 for round- and square-sectioned silica [60].

area of refractory. This results in the slag creeping up in the flow (Figure 1.11). In other words, if gravity force is greater than surface tension due to Marangoni flow, then the slag film will flow back. This phenomenon is known as slag creeping.

1.6.4.2 *Mechanism of Erosion in Carbon/Oxide Refractories*

Oxide refractories are prone to thermal shock. Modern refractories have much better thermal shock resistance through the addition of graphite particles, which absorb much of the stress generated by the expansion on heating of oxide particles. Erosion/corrosion of these oxide/carbon refractories occurs by a different mechanism [60] to that for slag line attack. Figure 1.12 shows wetting and non-wetting conditions. At high temperatures, reactive wetting is the norm; i.e., wetting will occur when there is reaction such as

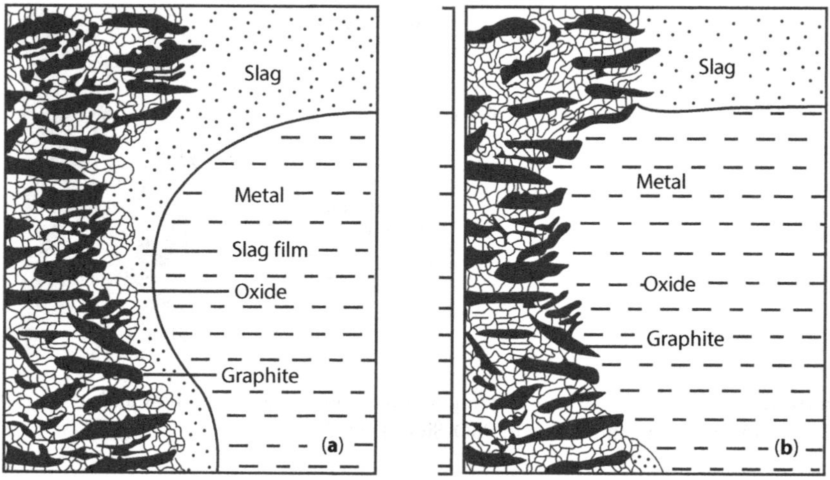

Figure 1.12 Schematic drawings showing the refractory erosion for oxide/carbon refractories [60] where the refractory surface is predominantly (a) oxide and (b) graphite, respectively.

i) oxide in the slag and ii) carbon in the metal. If the refractory surface at any one time consists mostly of oxides, the conditions will favor the covering of the surface by slag (since the oxide is non-wetting to the metal but slag reacts with and "wets" the oxide). Slag will then gradually dissolve these oxide particles (Figure 1.12). When the dissolution process results in a surface consisting of mostly graphite flakes the conditions become non-wetting to slag and the refractory is covered by metal since metal wets graphite (Figure 1.12). The dissolution of graphite by metal will occur until most of the surface is covered by oxide (Figure 1.12). At that point, the metal will retreat and the surface will be covered by slag once more and then another cycle of dissolution will occur.

In Figure 1.12, schematic drawings show the erosion of the oxide/carbon surface of refractories [60]. Refractory surface is predominantly (a) oxide and (b) graphite. Processes are adopted to reduce erosion. The erosion mechanism is briefly stated below:

(i) Erosion rates are directly related to flow rates of slag. Erosion rates can be slowed down by
  (a) Increasing viscosity of slag.
  (b) Reducing flow rates of metal (e.g., adopting electromagnetic braking on the liquid metal).

(ii) Driving force for erosion is the difference between saturation concentration ($C_{sat.}$) and actual concentrations (C_o) of carbon, e.g., ZrO_2 and ZrO_2/C refractories erosion can be decreased by adding 2% ZrO_2 to the slag to saturate it with ZrO_2 (for CC fluxes).

(iii) It is reported that making slag non-wetting extends the lifespan of refractories. The lifespan can also be extended by application of an electrical potential to the refractory and slag [61, 62].

(iv) ZrO_2 undergoes a phase transformation at elevated temperature transition and is therefore stabilized by CaO or Y_2O_3. However, CaF_2 is reported to attack stabilizers and promote erosion [63].

(v) Creation of a freeze lining on refractory surface. This is done by blowing the molten slag left in the vessel after draining the metal in order to produce a protective, consumable slag layer. Slag splashing is used in BOS practices. For successful slag splashing, FeO should be equal to 13% to get good adhesion with refractory and MgO % should be saturated (8%) to create a high-temperature phase, which slows down the erosion rate of the freeze lining [64].

1.6.5 Thermophysical Properties

Based on the previous discussion, it follows that thermophysical properties are essential. These are:

(i) Process control parameters that help solve industrial problems and improve product quality.

(ii) Mathematical models that can be worked out to enhance quality of products.

Models involve simulation of fluid flow and heat transfer during the process. This allows reliable predictions by generating accurate modulating input data and process variables such as viscosity, density, and thermal conductivity of the slag [65]. However, measurements are tricky because of high-temperature slag/container reactions. Considering a large number of processes involving many input variables is a massive task. This situation is aggravated by the fact that compositions of reactants frequently vary. In mathematical modeling and sensitivity analysis this problem can be eased by indicating which property data are more rational and the accuracy needed in such measurements. It is obvious that large bodies of property values are needed for a wide range of slag compositions. Given the difficulties of making these measurements, it follows that what we need are models and routines which will calculate reliable and accurate property values from chemical composition, since this is available on a routine basis for most processes.

1.6.5.1 Estimation of Thermophysical Properties

Different approaches are taken to estimate thermophysical properties. For this, a numerical analysis is made of the properties, e.g., correlation between liquidus temperature (T_{liq}) and chemical composition. More complex models relate property with the structure of slag for the case of T_{liq}. Recently, models based on neural networks have been reported. Following points same to be noted. For such studies, a binary phase diagram is often related and thermodynamic data for the binary system are exposed by the mathematical equation.

- Accuracy of model predictions of properties can never be greater than that of the database. If the data are based on measurements from a single laboratory and contain a systematic error then predicted values will also contain systematic error.
- There are discrepancies observed in properties measured in different laboratories. Thus, a database based on such measurements will tend to have a more scattered result but will be applicable to a wider range of slags. The optimum condition is to use a database based on assessed property data.
- Models based on the structure of the slag use one of the following:
  (i) Parameters such as (NBO/T) or the optical basicity (Λ) calculated from chemical composition.
  (ii) Thermodynamic properties relating to free energy change ΔGxs for stability of structure (i.e., bonding of strength).

1.7 Structure of Slags and Glasses

The structure of SiO_2 is a 3D array (Figure 1.13a,b) with (i) each Si^{4+} surrounded by 4 O^{2-} arranged tetrahedrally and (ii) each O^{2-} connected to two O^{2-} (denoted as bridging O) which causes the formation of a 3D array. Cations, e.g., Na^+ or Ca^{2+}, break the chain by forming a bond, O^{2-} -- Na^+ (denoted as nonbridging O). Slags and glasses are polymers in the form of chains, rings, etc., made up of SiO_4 4-tetrahedral units (Figures 1.13, 1.14). They have the following characteristics [66–69].

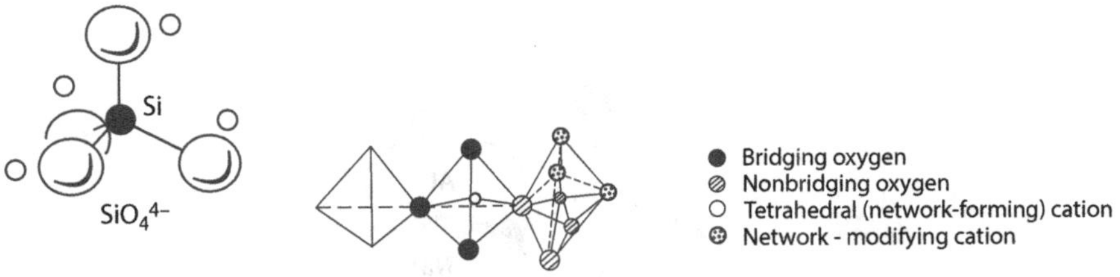

Figure 1.13 Structure of SiO_2 showing 3D characteristics [70].

Each Si (with a valence of 4) is surrounded (tetrahedrally) by 4 O^{2-} ions (with a valence of 2), each connecting to 2Si ions (Figure 1.13). In SiO_2 these SiO_4^{4-} polyhedra are connected in a 3D polymerized structure (Figure 1.13 and 1.14) and the oxygens are predominantly *bridging oxygens* (denoted O^o). Cations, such as Ca^{2+}, Mg^{2+}, etc., tend to break up the Si-O bonds and depolymerize the melt by forming *non-bridging oxygens* (NBO denoted O^-) and *free oxygens* (denoted O^{2-}), i.e., not bound to Si at all but to Na^+, etc. (Figure 1.14). Other cations, such as Al^{3+}, P^{5+}, and Ti^{4+}, can fit into the Si polymeric chain but need to maintain charge balance, e.g., if an Al^{3+} is incorporated into a Si^{4+} chain it must have a Na^+ (or one half of a Ca^{2+}) sitting near the Al^{3+} to maintain local *charge balance* (Figure 1.15). In practice, Ti^{4+} has been reported (on the basis of viscosity measurements [82]) to act as a network breaker (i.e., is in 6-fold coordination) and has been treated as such in this work. Smaller cations, such as Mg^{2+}, tend to give a wider distribution of chain lengths than larger cations such as Ba^{2+}. Cations, such as Fe^{3+}, in small concentrations can act as network breakers but in higher concentrations can be incorporated into the chain in a similar way to Al^{3+}.

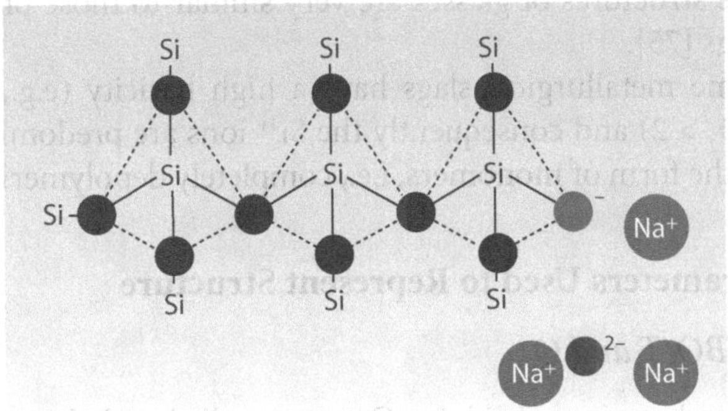

Figure 1.14 Schematic drawings and silicate chain with bridging O (O^0) shown in black, non-bridging O (O^-) as free O^{2-} as green [70].

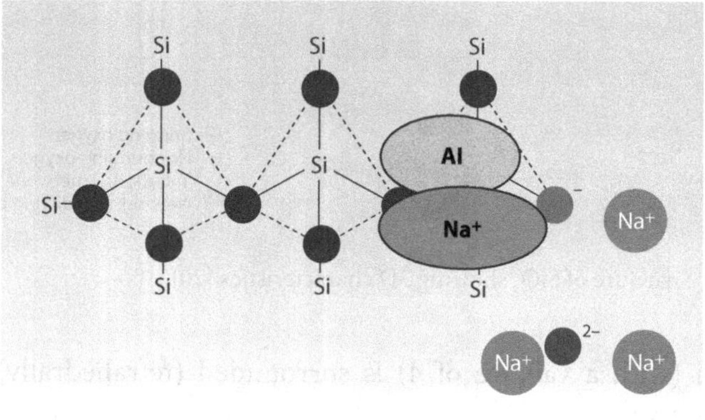

Figure 1.15 Schematic drawing showing Al^{3+} incorporated into silicate (Si^{4+}) chain, which requires cation (shown here as Na+) to maintain charge balance [70].

- The degree of polymerization can be expressed in terms of the numbers of bridging (NO^o), nonbridging (NO^-) and free oxygens (NO^{2-}). However, the degree of depolymerization is frequently represented by parameters such as the ratio of NBO/T (the ratio of the number of non-bridging O atoms to O atoms in tetragonal co-ordination), which is denoted as (NBO/T) [66-69] or the optical basicity (Λ) [69].
- The structure of melts (both slags and metals) can be represented using thermodynamic quantities (e.g., excess free energy [71, 72]) since thermodynamics provides a description of bond strengths.
- B^{3+} has 3-fold coordination in borosilicates but the structure of borosilicates (used as enamels) is a complex mixture of 3- and 4-co-ordination [73, 74].
- The structures of glasses are very similar to those of liquid slags [75].
- Some metallurgical slags have a high basicity (e.g., CaO/SiO_2 > 2) and consequently the Si^{4+} ions are predominantly in the form of monomers, i.e., completely depolymerized.

1.7.1 Parameters Used to Represent Structure

1.7.1.1 NBO/T and Q

NBO/T = number of non-bridging O/tetragonally bonded oxygen. NBO/T is a measure of depolymerization and is calculated using the procedures listed below. Cations like Na^+ or Ca^{2+} act as network breakers but if an Al^{3+}

goes into the chain it requires a cation (e.g., Na^{+}) to be sited near Al^{3+} to provide charge balance (Figure 1.15). A cation on charge-balancing duties cannot be used as a network breaker. The parameter $(NBO/T)_{corr}$ is a measure of depolymerization of slag. $(NBO/T)_{corr}$ can be calculated in the following manner, where f = fraction of Fe^{3+} ions acting as a network breaker (Eq. 1.9 and 1.10):

1) Calculate

$$Y_{NB} = \Sigma 2\left[\begin{matrix} X_{CaO} + X_{MgO} + X_{FeO} + X_{MnO} + X_{CaO} + 2X_{TiO2} + X_{Na2O} \\ + X_{K2O} + 3fX_{Fe2O} - 2X_{Al2O3} - 2(1-f)X_{Fe2O3} \end{matrix}\right] \quad (1.9)$$

Then calculate

$$X_T = \Sigma X_{SiO2} + 2X_{Al2O3} + 2fX_{Fe2O3} + \quad (1.10)$$

2)

$$\left(\frac{NBO}{T}\right)_{corr} = \frac{Y_{NB}}{X_T}. \quad (1.11)$$

3) Q is a measure of the polymerization of the slag and for this reason is often preferred to (NBO/T)

$$Q = 4 - \frac{NBO}{T} + \quad (1.12)$$

1.7.1.2 *Optical Basicity (Λ)*

One problem encountered for both (NBO/T) and Q is that they do not differentiate between the effects of different cations on the silicate structure. Optical basicity (Λ) has been used as a measure of structure principally because it provides some measure of differentiating between cations. It is calculated in the following manner (Eq. 1.13):

1) $\Lambda = \Sigma\,(X_1 n_1 \Lambda_1 + X_2 n_2 \Lambda_2 + X_3 n_3 \Lambda_3 + \ldots) / \Sigma(X_1 n_1 + X_2 n_2 + X_3 n_3 + \ldots)$ (1.13)

where n = number of oxygens in oxide (e.g., n = 2 for SiO_2 or 3 for Al_2O_3)

2) Λ^{corr} is calculated by deducting the mole fraction of cation required to charge balance the Al^{3+} in the chain. This is usually deducted from the largest cation present, e.g., Ba^{2+}, thus $X^{corr}{}_{BaO} = (X_{BaO} - X_{Al2O3})$

The following values of optical basicity are recommended:

Al_2O_3= 0.60: B_2O_3= 0.42: BaO= 1.15: CaO= 1.0: FeO= 1.0: Fe_2O_3= 0.75: K_2O= 1.4; Li_2O= 1.0: MgO=0.78: MnO= 1.0: Na_2O= 1.15: P_2O_5= 0.40: SiO_2= 0.48: SrO= 1.10: TiO_2= 0.61.

Various Λ values have been cited for CaF_2 0.43, 0.67 but a value of 1.2 should be used for viscosity calculations. However, recent work attributed to Sasaki *et al.* [76] indicates that Ca^{2+} bonds with $2F^-$ and thereafter is "out of the game" (thus, one can correct the slag composition by ignoring F concentration and deducting $2X_F$ from X_{CaO}) [77].

1.7.2 Structural Changes in Glasses and Slags with Temperature

When glass or a glassy slag (Q > 2.5) is heated C_p, thermal expansion and other properties show a marked change at a specific temperature (Figure 1.16). This temperature is called the glass transition temperature (Tg). Change in C_p and α is due to glass transforming into a supercooled liquid. This supercooled liquid (scl) exists until slag melts into a true liquid. However, in some slags in which Q < 2, supercooled liquid undergoes crystallization. Such transition is exothermic in nature. Upon heating, properties in glassy slag changes at T_g. These are described below:

(i) On an increasing C_p, C_p of glass is higher than crystalline slag over the entire range of temperature T_g and T_{liq}.
(ii) A three-fold increase in thermal expansion.
(iii) A remarkable decrease in viscosity with increasing temperature.
(iv) An abrupt decrease in slope of k as a function of temperature.

Viscosity (η) on glassy samples (Q > 2.5) are carried out. Cooling shows higher viscosities (Q = 2) and (ii) a smooth change with temperature as the liquid transforms to scl (with $\eta = 10^{12.4}$ Pas at T_g). In Figure 1.17b, viscosities

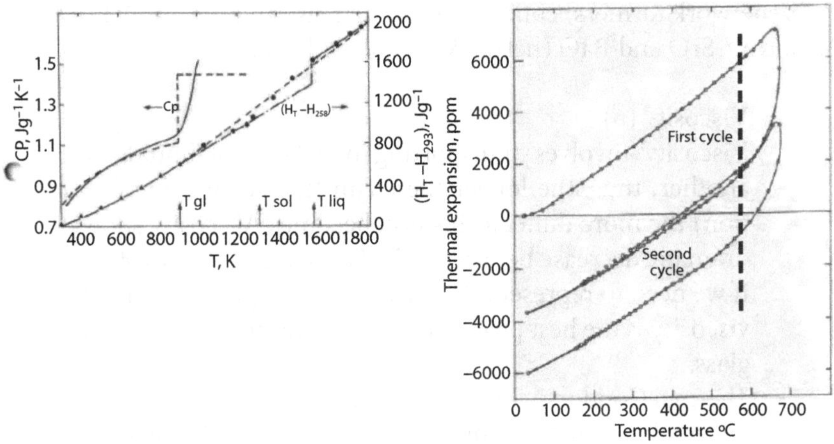

Figure 1.16 Heat capacity for a glassy slag (a) and thermal expansion coefficient as a function of temperature (b) [70].

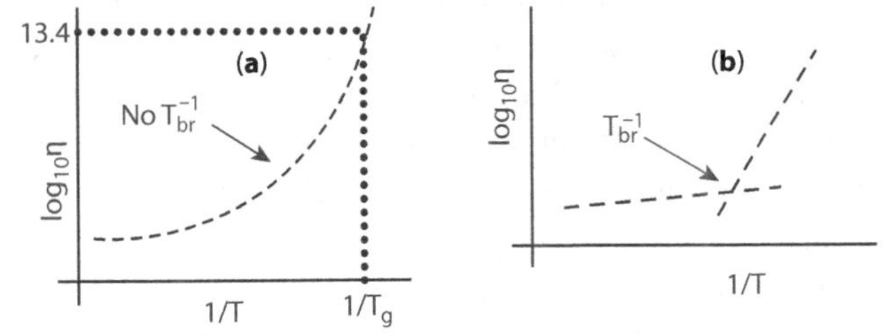

Figure 1.17 Plots of log10 viscosity as functions for reciprocal temperature (K-1) for (a) glassy slag (Q > 2.5) and (b) for a less glassy slag (Q = 2) [70].

for less polymerized slags (Q = <2) show a sharp change in slope at the break temperature, T_{br}. This indicates the first precipitation of crystalline materials. It can be noted that T_{br} decreases as the cooling rate increases.

1.7.3 Effect of Structure on Properties

The effects of polymerization or Q on properties are arranged, and properties showing the biggest dependence on structure are first, with a diminishing dependence on structure as we go down.

1.7.3.1 Effect of Structure (Q or (NBO/T)) on Properties

Property values for slags and glasses are greatly affected by the degree of polymerization (i.e., Q); polymerization (or Q) increases as (i) SiO_2 and

Al_2O_3 (network formers) concentrations increase and (ii) Na_2O, Li_2O, CaO, MgO, K_2O, SrO and BaO (network breakers) decrease.

- Viscosity (η)
  Viscosity involves transporting one layer of liquids over another; thus, the longer the chain length (or polymerization) the more difficult this task becomes. Viscosity shows a dramatic decrease between NBO/T = 0 and 1 (or Q = 4 to 3). If we need to represent the structure through a property, the viscosity is the best property to represent the structure of the glass.
- Thermal Conductivity (K) and Electrical Conductivity (K)
  Thermal conductivity of liquid slags tends to increase with increasing SiO_2 content (Figure 1.18c). Thermal conductivity (k) decreases with increasing NBO/T or decreasing Q. Thermal conductivity is more for covalently bonded Si^{4+} and Al^{3+} ions in the chain. Thermal conductivity is much lower for NBO and cations at the end of chains. This is expressed as easy conduction along the chain. Transport of phonons is much more difficult for chain-to-chain conduction. The thermal conductivity seems to be affected by the nature of the cations with $k_{Li2O} > k_{Na2O} > k_{K2O}$.
  Electrical conductivity involves movement of cations through a silicate network when an electrical field is applied. Thus, a well-developed network will hinder the movement

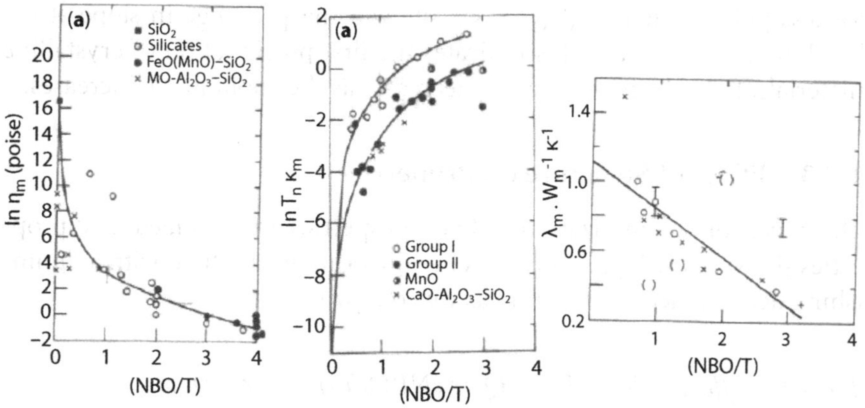

Figure 1.18 Viscosity (a), electrical conductivity (b) and thermal conductivity as a function of (NBO/T) of liquid slag at T_{liq} [70].

of these cations. electrical conductivity increases sharply by NBO/T increases or Q decreases and Mg^{2+}, Ca^{2+}, Sr^{2+} and Ba^{2+} are replaced by Li^+, Na^+, or K^+. Consequently, there are *two factors* affecting electrical conductivity:

(i) Degree of polymerization of slag with increases in electrical resistivity or decreases in electrical conductivity.

(ii) Size and concentration of cations since the number of cations slipping through the "holes" in the network will be greater for high concentrations of cations and for smaller cations.

- Thermal Expansion (α, β) and Density (ρ)
  Because of tight bonding, SiO_2 has a very low thermal expansion coefficient. Thermal expansion coefficient with first-order properties, i.e., $\alpha \equiv d\rho/dT$ (Figure 1.16a), tends to be more sensitive to temperature than the property itself (e.g., density). Density (ρ) is only slightly sensitive to polymerization (Q) in the slag. Thermal expansion increases with increasing NBO/T or decreasing Q. Thermal expansion is greater in M_2O if compared with MO and increases with increasing cation size (K > Na > Li and Ba > Ca > Mg). There is a marked change in thermal expansion when $T > T_g$ (ca. 300%), which occurs when glass changes into a suppercooled liquid (scl). However, accurate values of thermal expansion of glassy slags are difficult to obtain for T >Tg because supercooled liquid tends to collapse under.
- Surface Tension (γ) and Emissivity (ε)
  These properties predominantly depend on surface and not on the bulk. In mixtures, constituents with lowest *surface tension* will tend to occupy the surface layer (e.g., sulfur in steels and surfactants like B_2O_3 in slags). Surface concentration will depend upon both the surface tension and chemical activity of the component (e.g., B_2O_3 in the slag). When there are two or more surface-active components in the slag (e.g., CaF_2 and B_2O_3) there will be competition for the surface sites. The surface tension is a thermodynamic propert y. Surface tensions of some oxides are given in Table 1.4. Since $\gamma SiO_2 > \gamma$ for other oxides, it follows that γ will decrease with increasing SiO_2 and increasing Q.
  Slags and glasses are semitransparent to IR radiation. Slags and glasses are optically thick if $\alpha^*d > 3$, where α^* = absorption coefficient and d = depth. In practice, this means

Table 1.4 Surface tensions (mNm^{-1}) of pure slag components at 1773 K [70].

C	SiO_2	CaO	BaO	SrO	MgO	$A_{l2}O_3$	FeO	NiO	MnO	CrO	Na_2O	K_2O	TiO_2	ZrO_2	Cr_2O_3	Fe_2O_3	CaF_2	B_2O_3
γ	260	625	560	600	635	655	645	645	645	360	295	160	360	400	800	300	290	110

*Note: C = Components of pure slag and γ = Surface tension of pure slag *emissivity* of liquid slags will depend on the depth of the slag sample. Values are usually cited for the *optically thick condition.*

- Heat Capacity (C_p) and Entropy of Fusion (ΔS^{fus})
  These properties are little affected by structure and Eq. 1.14 is valid, where 1, 2, etc., are subscripts representing different components. However, C_p is dependent upon the nature of slag and C_p jumps abruptly at T_g when a glass changes to a supercooled liquid (scl), whereas a crystalline slag shows no jump at T_g (2) $C_{p\,scl} > C_{p\,cryst}$ over the range T_g to T_{liq} and (3) exhibits an enthalpy of fusion (ΔH^{fus}) at T_{liq}, whereas a supercooled liquid exhibits no ΔH^{fus}.

$$P = \Sigma\left[\left(X_1\overline{P_1}\right)+\left(X_2\overline{P_2}\right)+\left(X_3\overline{P_3}\right)+\left(X_4\overline{P_4}\right)\right] \tag{1.14}$$

1.7.4 Factors Affecting Property

Electrical resistivity increases with increasing size of polymer units. Thus, electrical resistivity increases with increasing Q and decreases with a decrease in NBO/T and optical basicity of material. Since addition of Al_2O_3 effectively increases polymerization, electrical resistivity will be a function of Qcorr, (NBO/T)corr, and optical basicity (Λcorr). The nature of cations also affects electrical conductivity. The charge (z) on ions affects the resistance, since a Ca^{2+} ion will bridge 2 chains whereas Na^+ will not (Figure 1.19). Therefore, resistance to the movement of ions will be greater in CaO-bearing slags than Na_2O-bearing slags. However, other factors are:

(i) There are twice as many Na+ cations to Ca^{2+} (for any given Q), i.e., twice the number of potential cations available for transport.
(ii) Smaller cations will slip through holes in the network more easily than bigger cations. This is offset by the fact that a smaller cation has higher polarizability.

$O^- - Ca^{2+} - O^- \quad O^- - Na^+ \quad Na^+ - O^-$

Figure 1.19 Schematic diagram showing (a) "bridging" of two O^- in chains by Ca^{2+} and (b) absence of bridging with Na^+ ions [70].

Zhang *et al.* [78] have reported M_2O–SiO_2 bearing slags. In this slag, electrical conductivity (κ) decreases with cation radius (r), whereas κ is reversed for Mg, Ca, Sr, and Ba oxide-slags. It is considered that all cations engaged in charge-balancing duties are not participated in their movement (i.e., electrical conduction).

Zhang *et al.* [78, 79] have plotted ln κ vs ln η (Figure 1.20). They found that for a given viscosity, electrical conductivity of monovalent cation, i.e., $\kappa(M^+)$, is greater than electrical conductivity of divalent cation, i.e., $\kappa(M^{2+})$. This is attributed to a greater concentration of M^+ ions compared to M^{2+} ions and a greater number of bridging chains (see below).

Transition metal oxides have more than one valence (e.g., Fe^{2+} and Fe^{3+}). Consequently, ionic conduction can occur by charge transfer between Fe^{2+} and Fe^{3+}.

$$Fe^{3+} \rightarrow Fe^{2+} + e$$

Similar reactions can occur with other transition metal oxides (e.g., Cr^{3+} and Cr^{2+}). At high concentrations of FeOx, etc., there is the possibility of electronic transport, which causes a significant increase in the total conductivity [80]. Electrical conductivity of slags containing FeOx are determined as a function of p_{O2} and κ, which is a maxima at Ca $p_{O2} = 10^{-6}$atm [81]. An attempt has been made by Jahanshahi *et al.* to model the oxidation state relating to electrical conductivity [81, 82].

1.7.4.1 *Size of Cations*

The size (r) of the cations affects the conductivity because small cations are more mobile than larger cations and can slip through the "holes" in the network. Zhang *et al.* [78, 79] plotted data in Figure 1.20 for Ba^{2+}, Sr^{2+}, Ca^{2+}, and Mg^{2+}, in which it can be seen that for a given viscosity the conductivity

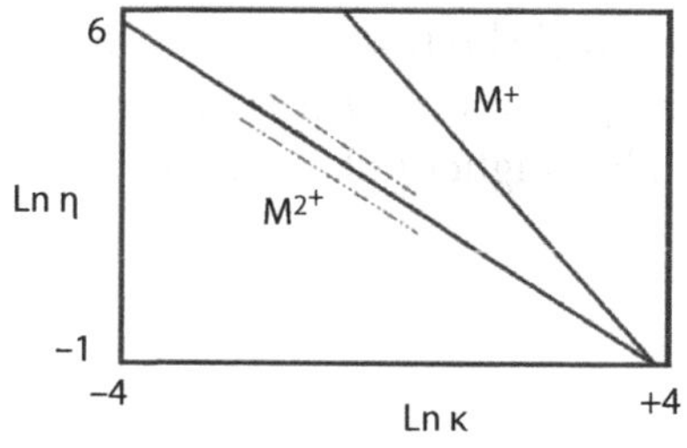

Figure 1.20 Schematic diagram representing plot of ln (viscosity, Pas) vs. (ln conductivity, Ω^{-1}. m^{-1}) for divalent and monovalent ions (after Chou); dash-dot line = Ca^{2+}; dash-double dot line = Ba^{2+} [79].

of $Ca^{2+}$and Mg^{2+} is slightly greater than Ba^{2+}; this may be due to the greater mobility of the smaller cation or may be due to greater resistance to "bridging of chains" with bigger cations.

1.7.5 Factors Affecting Measurements

Polarization is the principal problem that occurs due to concentration gradients arising in the electrolyte and manifests as a back EMF opposing the normal ionic flow. The four-probe method is used to minimize polarization, in which two electrodes carry the current and the other two monitor the potential. Frequency (f) has been reported to affect κ [83]. Some studies assume that the impedance of the electrolyte region is a function of f^{-1}.

1.7.6 Measurement Methods

There are various measurement methods described in the *Slag Atlas* [83] and, in general, it is preferable to have the two electrodes with similar areas (Figure 1.21c,d). These methods can be divided into the various 2-electrode (Figure 1.21a,b,c) and 4-electrode methods (Figure 1.22d), with preference being given to the latter since it minimizes problems associated with polarization.

1.7.6.1 *Methods of Estimating Electrical Conductivity*

There are no models available to calculate electrical conductivity (κ). Jiao and Themelis [84] have obtained an empirical relation for $CaO+MgO+MnO+SiO_2$ quaternary slags at 1773 K. However, the developed Eq. 1.15 is restricted to estimating electrical conductivity of industrial slags:

$$K_{1773} = -3.34 + 6.41X_{CaO} + 6.45X_{MgO} + 8.06X_{MnO} \qquad (1.15)$$

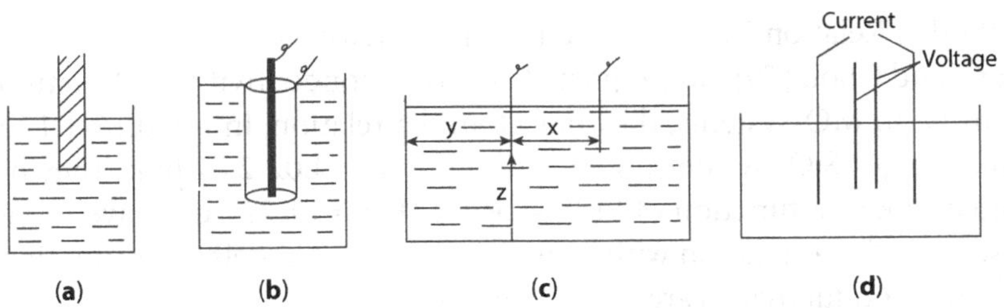

Figure 1.21 Cell arrangements for measuring electrical conductivities of molten slag (a) central electrode; (b) ring cell; (c) 2-electrode cell; and (d) 4-electrode cell [70].

However, Zhang and Chou [78, 79] have reported two models to calculate thermal conductivity of some silicate melts.

Method 1: Estimation of Electrical Conductivity Using Optical Basicity
Factors affecting electrical conductivity are:

- (i) Structure of network, which restricts the movement of cations
- (ii) Nature of cations
- (iii) Viscosity
- (iv) Size of cations
- (v) Charge on cations
- (vi) Polarizability of cations.

Zhang and Chou [78] have used corrected optical basicity (Λ^{corr}) to cover the above factors. The model is described below (Eqs. 1.16 and 1.17);

$$\ln k = \ln A_k - \left(\frac{B_k}{R^* T}\right) \tag{1.16}$$

$$B_k = m(\Lambda^{corr}) + n \tag{1.17}$$

Experimental data are regressed to obtain best fit values for Aκ, m and n as 5054, 3.24×10^5 and $-2.768.10^5$ respectively for the CaO-MgO-Al_2O_3-SiO_2 system; and 9462, 3.197×10^5 and -2.49×10^5 respectively for CaO-Al_2O_3-SiO_2 system. Results are plotted in Figure 1.21. It is possible to distinguish small deviations of individual cations (Ba to the left and Ca and Mg to the right of the curve). The mean uncertainty in the calculated values is less than (< 15%). The model can be applied to slags whose composition lay within the range $\Lambda^{corr} = 0.58$-0.67.

Method 2: Relation Between Electrical Resistivity and Viscosity
Zhang and Chou [79] have reported electrical conductivity as a function of viscosity. In MO-systems, they have found a relation; $\ln \eta = 0.15 - 1.1 \ln \kappa$. Also, in M_2O-SiO_2 systems, the relation is $\ln \eta = 4.02 - 2.87 \ln \kappa$. They have plotted ln κ as a function of ln (viscosity) (Figure 1.22) and have found a reasonably linear relation with some scatter. Thus, models for estimation of electrical conductivities are in their infancy.

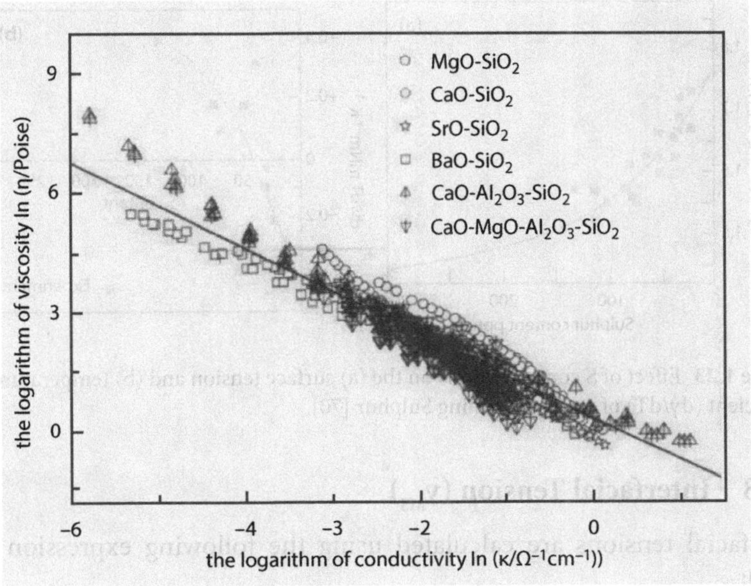

Figure 1.22 ln electrical conductivity as a function of ln viscosity [79].

1.7.7 Factors Affecting Surface Tension (γ)

Surface tension is not a bulk property. Surface-active components, such as sulfur or oxygen in molten metals, have a low surface tension as compared to metal. Consequently, it tends to occupy the surface layer in much higher concentrations than exist in the bulk. Thus, surface tension and its temperature dependence ($d\gamma/dT$) surface tension are extremely sensitive to sulfur or oxygen concentrations (Figure 1.23a). In slags, there is not more difference between surface tensions of surface-active components (i.e., B_2O_3, CaF_2 and SiO_2-based slag components and CaO-based slag component), which is shown in Table 1.4. Surface tensions of slags are not structure-dependent. But surface tension decreases with increasing Q value (example: $\gamma_{SiO2} < \gamma_{CaO}$) and is shown in Figure 1.23b.

Surface tension of slag is dependent upon concentration of surface active components and chemical activity of the surface active component. So, models of surface tension tend to be based on thermodynamics of slag system.

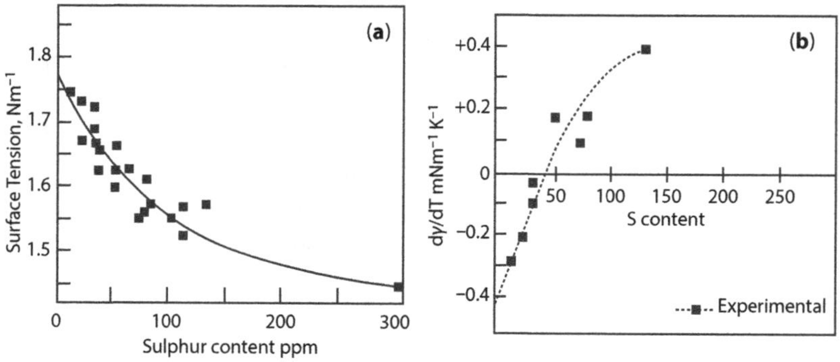

Figure 1.23 Effect of S concentrations on the (a) surface tension and (b) temperature coefficient (dγ/dT) of steels containing Sulphur [70].

1.7.8 Interfacial Tension (γ_{MS})

Interfacial tensions are calculated using the following expression (Eq. 1.18);

$$\gamma_{MS} = \gamma_M + \gamma_S + 2\varphi(\gamma_M . \gamma_S)^{0.5} \tag{1.18}$$

Where ϕ is an interaction coefficient. The parameter ϕ is found to have a value of 0.5 for slags free of FeO and increases with addition of FeO [85]. Tanaka *et al.* [72] proposed the equation given below (Eq. 1.19);

$$\phi = 0.5 + 0.3X_{FeO} \tag{1.19}$$

Qiao *et al.* [86] have sketched a model for estimating interfacial tension using excess interfacial tension ($\gamma^{xs}{}_{MS}$). Values are obtained from fitting experimental data (Figure 1.24).

1.7.9 Factors Affecting Thermal Conductivity

Thermal conductivity of slags [87–96] involves movement of phonons. This is usually termed as phonon conduction. Therefore, thermal conductivity (k) depends on structure. Thermal conductivity of molten slags increases with increasing SiO_2 content in the slag. Thermal resistance (1/k) is the backbone of phonon transport. Phonon transport is small when passing along the chain if compared to phonon transport between the chains. Therefore, more slag is polymerized (i.e., Q increases or NBO/T

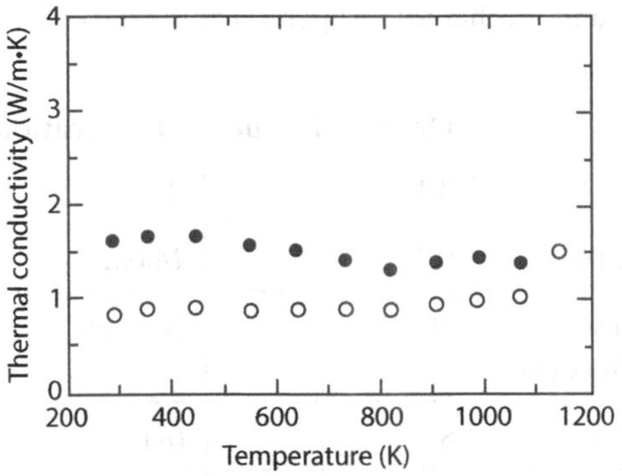

Figure 1.24 Thermal conductivity of glassy (O) and crystalline (•) slags as a function of temperature [70].

decreases), and the thermal conductivity is higher (Figure 1.24). This signifies that thermal conductivity depends on structure.

(a) Thermal conductivity is a function of viscosity [87].
(b) Thermal conductivity (k) is also a function of temperature (i.e., Vogel-Fulcher relation and Arrhenius equation). From the thermal conductivity data, it is concluded that cations affect thermal conductivity values for alkali slags M_2O- SiO_2.

1.7.9.1 *Thermal Conductivity of Crystalline Slags*

Nishioka *et al.* [96] have measured thermal conductivity of CaO + Al_2O_3 + SiO_2 ternary slags system in glassy and glassy state. The value of the thermal conductivity of crystalline and glassy state of the above ternary system was found to be k_{cryst} = (1.5 to 2.5) k_{glass}. This is due to the increase in packing densities. Figure 1.24 shows temperature coefficient (dk/dT), which is independent of temperature (200–1200 °C).

1.8 Physical and Mechanical Properties of the Ferrochrome Slag Aggregate

Water absorption and specific gravity tests were carried out in accordance with IS 2386 Part III-1963 [98]. Shape tests were performed according to IS 2386 Part I-1963. Crushing value, impact value, and abrasion value

Table 1.5 Physical and mechanical properties of the ferrochrome slag aggregate [97].

Test performed	Observed value	Specification as per MoRTH
Specific gravity	3.16	2.5-3
Water absorption (%)	0.37	Max.2
Combined flakiness and elongation indices (%)	20.37	Max.35
Angularity number	8	0-11
Impact value (%)	9.63	Max.30
Crushing Value (%)	22.75	Max.45
Abrasion value (%)	22.39	Max.40
Density in loose state (g/cc)	1.661	---
Density in dense state (g/cc)	1.792	---

tests were performed according to IS 2386 Part IV-1963. The physical and mechanical properties of the ferrochrome are given in Table 1.5.

1.9 Characterizations of Ferrochrome (FeCr) Slag

Ferrochrome slag is characterized to find out leachability of chromium from the slag. The following characterization studies are carried out:

- Evaluation of physical properties;
- Chemical characterization of slag;
- Mineralogical studies for assessing presence of chromium-bearing mineral phases and other phases of slag;
- Microstructure characterization for evaluating structure of phases present in slag.

1.9.1 Physical Characterization

Evaluation of size, colour, specific gravity, etc., of ferrochrome slag.

Experimental Methods:

- Particle Size Distribution (PSD) Analysis of FECR Slag
  Particle size distribution (PSD) analysis is an important requirement for leachability study as well as from utilization aspects. PSD of granulated FeCr slag and lumped slag are determined by standard sieve analysis as stipulated in IS 2386 (Part I)-1997 [99]. The procedure consists of gently placing the material to be sieved on the test sieve having specified nominal aperture size and separating the material by shaking, tapping or washing into oversize, and undersize. A sieve test apparatus is shown in Figure 1.25. 600 gm of granulated slag sample is kept on the top of 4750 micron as shown in the Figure 1.25. Standard sieve sizes chosen are 4750, 2360, 1180, 600, 300, 150, 75 microns for the purpose. Mass retained on each sieve is collected and weighed accurately. Cumulative mass passed and cumulative mass passed percentage through each sieve are estimated. From this

Figure 1.25 Sieve test apparatus [100].

experiment, average particle size is determined. It is found in the range of 8–20 mm size.

- Specific Gravity and Water Absorption
Specific gravity of coarse aggregate is determined as per the procedure outlined in IS 2386 (Part I)-1997 [99]. Two grams of sample is thoroughly washed to remove finer particles and dust, drained. These are then placed in the wire basket. The basket is immersed in distilled water at a temperature between 22 °C and 32 °C with a cover of at least 5 cm of water up to the top of the basket. Immediately after immersion, the entrapped air is removed from the sample by lifting the basket containing it, 25 mm above the base of the tank, and allowing it to drop 25 times at the rate of about one drop per second. The basket and aggregate remain completely immersed during the operation for a period of 24 hours. The basket containing sample is then jolted and weighed in water. If it is necessary, then it is to be transferred to a different tank for weighing purpose. They are jolted 25 times as described above in the new tank before weighing (weight A1). The basket containing aggregate then is removed from water and is allowed to drain for a few minutes, after the aggregate is gently emptied from the basket onto one of the dry clothes. The empty basket is again placed in water, jolted 25 times and weighed in water (weight A2). The aggregate placed on the dry cloth is gently surface dried with the cloth, transferring it to the second dry cloth when the first is removed to ensure that there is no moisture. Aggregate is then placed in the oven in the shallow tray and is transferred to an oven. Temperature is maintained between 100 °C and 110 °C for 24 h. It is then removed from the oven, cooled in the airtight container and weighed (weight C).

$$Specific\ gravity = \frac{C-B}{A} \quad \text{..............................(1)}$$

$$Where\ A = A_1 - A_2$$

$$Water\ absorption(percent\ of\ dry\ weight) = \frac{100(B-C)}{C}$$

- Specific Gravity of Granulated Ferrochrome Slag
  A representative sample of 500 gm of the above saturated surface dry material is obtained. Sample is kept immersed in distilled water for 24 hours. Then the sample is dried. This weight of saturated and surface dry material is recorded as A. Then the slag sample is placed in a pycnometer and is filled with distilled water to remove any froth from the surface. The pycnometer with sample is weighed as B. The pycnometer is then emptied and is filled with distilled water to the same level as before. The pycnometer with water is weighed as C. The sample is then oven dried at temperature of 100–110 °C and weighed accurately as D. The relation used for finding out specific gravity is given below:

$$Surface\ dry\ basis = \frac{D}{A-(B-C)} \text{..............................(1)}$$

$$Water\ adsorption(\%\ dry\ weight) = \frac{100(A-D)}{D} \text{..............................(2)}$$

Where

A = *Weight of saturated dry sample*
B = *Weight of Pycnometer with sample and water*
C = *Weight of Pycnometer with water*
D = *Weight of Oven dried sample*

1.10 Environmental Aspects of Ferrochrome Production

Technologists are concerned about the environmental impact of their actions. Ferrochrome producers are no exception, particularly due to the release of hexavalent chromium into the atmosphere, water and soil. Hexavalent chromium [Cr(VI)] also poses a threat to human and animal life. Therefore, FeCr production should be managed so as to have no adverse effects on the environment. Judicial control of FeCr production is mandatory to protect human and animal life. Currently, established techniques are available to combat noise pollution and to prevent pollution of the ambient air. Chromium in hexagonal form results from the production of ferrochromium, with Cr(VI) being a suspected carcinogen. In a 1991 DECD Report entitled "The State of the Environment," chromium is associated with lung cancer, gastrointestinal cancer and dermatitis [101]. However, even though chromium is an element needed by man—the difference lies in the oxidation state and intake. According to the US Department of Health and Human Services, a daily intake of 50 to 200 pg Cr(Ill) is recommended for adults to assist in the metabolism of sugars, proteins and fat [102]. On the other hand, exposure to high concentrations of Cr(VI) can be lethal—a lethal ingestion of Cr(VI) has been stated as 7,5 mg/kg body mass [103].

Production of high-carbon ferrochromium (HCFeCr) and charge chrome (ChCr) in an electric submerged arc furnace has the following environmental impact.

1.10.1 Raw Material Preparation

Handling, crushing, screening and proportioning of chrome ore do not present any health or environmental risk. Noise and dust pollution associated with the above activities do not cause any major health issues. However, in order to determine the hazardous effect of hexavalent chromium, a representative sample of chrome ore concentrate (typically -0,5 mm with a CrZ03 content of 45%) was analyzed. The Toxicity Characteristic Leaching Procedure (TCLP) specified by the United States Environmental Protection Agency (US EPA) was followed (in accordance with requirements stipulated in the Federal Register, Vol 51, November 1986). TCLP extract from the sample contained 0,12 ppm total chromium and $<$ 0,1 ppm Cr(VI). These levels are below the limit of 5 ppm Cr imposed by the US EPA, which implies that chrome ores and concentrates are not classified as a hazardous substance with respect to chromium [104].

1.10.1.1 Smelting

Solid products obtained from a smelting furnace are metal, slag and dust. In open-top furnaces, dust is collected in a bag filter plant, whereas in a closed top furnace, dust is scrubbed in a venturi system and is produced as slurry. Chromium in its metallic state (zero valency) as ferrochrome is generally considered harmless as per norms prescribed by the International Agency for Research on Cancer (IARC) (monograph titled *Evaluation of Carcinogenic Risks for Humans*, Volume 49) where both metallic chromium and Cr(Ill) are classified as Group 3 compounds, i.e., these compounds are not classifiable as to their carcinogenicity to humans [105].

No special precautions regarding toxicity are thus necessary when storing, transporting or using the alloy. Slag is also regarded as harmless in terms of the IARC classification, as the chromium is predominantly present in ferrochrome slags as Cr(Ill). As such, in South Africa it can safely be consigned to a Class II landfill for nonhazardous substances with less onerous conditions than for a Class I landfill. However, there exists a great deal of controversy regarding the possibility of oxidation of Cr(Ill) to Cr(VI). Numerous papers have been published on this topic; none of which relate specifically to ferrochrome slags in landfills. Rather, the papers under review have addressed the issue of Cr(Ill) oxidation under laboratory conditions using Cr(Ill) solutions in the presence of strong oxidants such as MnOz. In order to allay any fears that may exist regarding the possibility of Cr(Ill) in ferrochrome slags oxidizing to Cr(VI) in a landfill, and thus slowly releasing Cr(VI) into the environment, carefully controlled experiments are set up to simulate conditions existing in a typical ferrochrome slag dump in South Africa. These experiments are scheduled to run over a number of months, but results to date indicate that there are no hazards associated with ferrochrome slag dumps in terms of the potential to leach hexavalent chromium. The greatest risk to health and the environment in the smelting of ferrochrome lies in the dust from the bag filter or sludge from the gas cleaning system. This is particularly the case for open-top furnaces where chromium is readily oxidized to Cr(VI) in the off-gas dust. Hexavalent chromium also occurs in the sludge and circulating water from the gas scrubbers of closed-top furnaces, albeit to a much lesser extent. Another factor which influences the occurrence of Cr(VI) in the off-gas dust stream is the chemical composition of the slag. Furnaces operating under acidic slag regimes have significantly lower water-soluble Cr(VI) content of the dust than furnaces associated with basic slag practice. Table 1.6 illustrates this trend.

Table 1.6 Water-soluble Cr(IV) content of furnace dust.

Closed Furnace	Acid slag practice	5
	Basic slag practice	100
Open Furnace	Acid slag practice	1000
	Basic slag practice	7000

1.10.2 Environmental Management Program

Numerous options are available to counteract and prevent water and soil pollution. As with most problems, it is preferable to treat the cause and not the symptoms. Furthermore, a piecemeal approach to any pollution problem will not be cost-effective. A holistic approach is required, which should culminate in the formulation and implementation of an Environmental Management Program (EMP). Various techniques can be used. The methodology adopted by Samancor has been found to be successful. At the outset, it is realized that specialist knowledge and skills will be required to produce the most cost-effective EMP. Services of a consulting engineering firm are to be retained to work in close collaboration with plant personnel.

1.10.2.1 Geohydrological Survey

The first step in a remediation program is to determine the extent of the problem, if any. At Samancor's Ferrometals chrome smelting works in Witbank in Mpumalanga, South Africa, the process involves a geohydrological study to determine soil and rock conditions in order to assess whether groundwater contamination is occurring. If so, a model hydrogeology of the site in the vicinity of potential sources of contamination needs to be worked out. The production facilities at Ferrometals consist of two closed-top charge chrome furnaces, an open ferrosilicon furnace, a bottom blown convertor for the production of either medium carbon ferrochromium or ferromanganese, and three open-top furnaces currently producing ferromanganese.

Investigations include an aerial survey for identifying geological faults and topology. Auger and core drilling is required to study ground formation as well as abstraction of groundwater from these boreholes to determine water quality. Another important reason for the geohydrological survey was to identify suitable areas for future landfills. A specific methodology was followed in accordance with "Guidelines for Geohydrological Investigations as Pertaining to Industrial Ground Water Pollution," July 1992, of the Department of Water Affairs and Forestry (DWAF).

A conceptual geohydrological model was proposed to identify both a confined aquifer as well as a perched aquifer [106]. Analysis of groundwater and slurries revealed that water quality is acceptable both in terms of total chromium as well as Cr(VI). High sulphate levels are attributed to former coal mining activities in the area. Evaluation of possible future waste disposal sites was conducted in accordance with guidelines relating to "Minimum Requirements for Waste Disposal Facilities" (DWAF, 1993).

The hydrogeological process involves a 6-step plan whereby sites are evaluated and ranked based on criteria such as logistics, topography, geohydrological properties and environmental factors. The South African Department of Mineral and Energy Affairs has produced an Aide Memoire for preparation of an environmental management programme report (EMPR). EMPRs are only required for activities falling under the former Mines and Works Act. This act did not apply to Ferrometals; nevertheless, the document provides valuable guidelines firstly to undertake an Environmental Impact Assessment (EIA) and secondly to formulate an Environmental Management Programme (EMP). The latter consists of the following phases: a construction phase, an operational phase, a decommissioning and closure phase. A timetable identifying the duration and sequence of each activity as well as the financial implications of the EMP form the last two phases. In the case of Ferrometals, the development of an EMP was relatively simple, as the results of the geohydrological investigation indicated that groundwater is not contaminated by Cr(VI). To control surface water, a series of canals and suitably engineered catchment dams are proposed to prevent run-off of storm water from the site. In addition, regular pumping and analysis of borehole water is proposed to monitor groundwater quality. A series of boreholes is drilled as part of the investigation to develop an EMP. The location and depth of these holes are planned to provide maximum information as to the hydrogeology of the site. Water quality from boreholes down to gradient from the effluent ponds is found to be within the legal limits. These effluent ponds contain sludge from the gas cleaning plants of the closed-top charge chrome furnaces.

1.10.3 Remediation

Remediation of polluted water and soil is of utmost importance. For greenfield sites, however, the task is much simpler. A proper environmental impact assessment (ElA) is required beforehand. As it is said, prevention is always better than cure.

1.10.3.1 Environmental Impact Assessment

The purpose of an environmental impact assessment (EIA) is to discover possible future activities that may affect the environment in accordance with current as well as pending legislation in order to recommend ways to minimize and control the impact of these activities. In South Africa, there are currently twelve acts applicable to the environment, the most important of which are the

(i) Environment Conservation Act
(ii) Water Act
(iii) Health Act
(iv) Atmospheric Pollution Prevention Act
(v) Hazardous Substances Act.

Similar situations will hold for other countries as well. The next logical step is the formulation of an EMP, as mentioned previously. However, the most important phase is the implementation of an ongoing commitment to the EMP.

1.10.3.2 Chemical Reduction of Cr(VI)

One of the remedies is to reduce Cr(VI) to Cr(Ill) by reduction. Since solid-state reduction is not feasible, the dust must be leached to get the water-soluble chromium into solution before reduction. The most commonly used reducing agent is ferrous sulphate (FeS04). As the price of this commodity is very high, over the last year in South Africa, Samancor launched an investigation to determine the optimum "window" in terms of La. pH of leachate, temperature, and reaction time. By controlling experiment parameters, the amount of FeSO4 has been added to minimize the addition, whilst at the same time ensuring that the resultant water quality conforms to the set standards [107–115].

1.10.3.3 Bacterial Reduction of Cr(VI)

In their study, Apel and Turick comprehensively discussed bioremediation of hexavalent chromium by bacterial reduction [107–115]. It is shown that the bacteria *Pseudomonas aeruginosa* and *Pseudomonas fluorescens* both have the ability to reduce Cr(VI) to Cr(Ill). However, Samancor operated closed-top charge chrome furnaces where a significant percentage of the carbon is in the form of coal. It resulted in a buildup of phenols in the closed-circuit scrubbing water of the venturi gas cleaning system.

A laboratory experiment is currently underway where Cr(VI) leachate is spiked with 50 ppm phenol and the bacteria introduced into solutions of Cr(VI)–phenol water together with nutrients. It was found that the bacteria can tolerate the presence of phenols. The next step is to determine whether the bacteria can utilize the phenols as an energy source, thereby solving the Cr(VI) and phenol problems simultaneously.

1.10.3.4 Solidification/Stabilization

This topic has also been researched extensively. However, no reference is found in regard to stabilization/solidification (S/S) of ferrochrome bag filter dust, and therefore it was necessary to undertake a S/S program specifically for bag filter dust [107–115]. The most commonly used S/S agent is Portland cement. A series of laboratory tests were conducted using Portland cement in different concentrations to solidify the dust and stabilize, i.e., the hexavalent chromium. Evaluation of the S/S products included leaching tests, permeability measurements and strength tests. Results have indicated that cement additions of between 7,5% and 15% has resulted in solidified masses, which has leached < 5 ppm Cr using the EPA TCLP. Furthermore, testing is underway to determine long-term stability and strength of the solidified dust using combinations of Portland cement, fly ash and lime. However, it has already been demonstrated that S/S techniques have the potential to prevent Cr(VI) contamination of soil and water from bag filter dust.

1.10.3.5 Other Methods of Remediation

There are numerous other methods available to prevent Cr(VI) in bag filter dust or scrubber sludge, thus protecting against environmental pollution. An obvious procedure is to recycle the dust back to the furnace. In order to do this, some form of prior agglomeration is necessary, as fine dust cannot be returned to the furnace as such. Well-proven methods of agglomeration include briquetting and pelletizing. However, these are costly options which require additions of binders as well as curing processes to render agglomerates strong enough to withstand handling and to prevent thermal spalling.

Samancor is researching the suitability of an agglomeration process which requires neither a binder nor a curing process, but relies on a high-intensity mixer. Results to date are promising. The long-term effect of recycling heavy metals back to the furnace must still be determined. Bleeding off a stream of dust from the bag filter at appropriate intervals is envisaged to prevent a buildup of such metals. Another technique is DC plasma arc melting/smelting of bag filter dust to produce an environmentally acceptable slag, whilst at the same time recovering residual chromium and iron oxides from the dust as ferrochrome. The plasma furnace is also a useful tool, i.e., to remelt unsaleable metal fines and chrome-rich scrap. Finally, it has been shown that selected grasses from other plants have the ability to take up chromium from the soil [107-115]. Unfortunately, the chromium taken up by plants remains mostly in the roots. Chromium removal by harvesting the stem and leaves of the plants thus does not seem feasible. Nevertheless, it affords a method of polishing Cr(VI) contaminated soils. Suitable plants for South African conditions have been identified and laboratory trials are underway.

1.11 Conclusions

Major consumers of ferrochrome alloy require 5 metric tons of ferrochrome per year. This is obtained from chromite mineral which belongs to the spinel group. The main products obtained from this mineral are ferrochrome alloy and metallic chromium. Production methods involve a smelting operation which need per tons of alloy produced. Thus, the process is highly energy-intensive. Revenue generation depends on chromium to iron ratio.

In this chapter, the four primary smelting processes currently available were discussed in detail. Production of ferrochrome alloy involves carbothermic reduction of chromite. During the process, carbon is picked up by the alloy. However, stainless-steel manufacturers require low-carbon ferrochrome alloy. Therefore, further processing is required to eliminate carbon in preference to chromium present in ferrochrome alloy.

Thermodynamics data suggest that process parameters like oxygen partial pressure and oxidation temperature can be controlled to eliminate carbon in preference to chromium. The above issues were discussed in detail and suitable diagrams, thermodynamics equations, and other data were presented.

Other production methods like silicothermic process and aluminothermic process of ferrochrome alloys were also discussed. The advantages of these processes are based on the fact that ferrochrome alloy is not contaminated with carbon. Chemical reactions are exothermic in nature; therefore, use of electric energy is limited and the processes are deemed to be advantageous.

Various technologies were also described in which the use of electric energy is reduced by using fine ores in the form of pellets by pre-reducing pellets.

Fundamental knowledge relating to chromium-bearing slags was reviewed and discussed, including the oxidation state of chromium, the activities of oxides, the phase diagrams, and factors affecting the properties of slag. The physical properties, viscosity and electrical conductivity of FeCr slag and their industrial applications were also discussed.

Environmental impact assessment studies proved that FeCr slag is a nonhazardous material and environmentally friendly, and therefore can be used as a green construction material in lieu of natural aggregates. Waste is a source of secondary raw materials but at the same time can negatively impact the natural environment and public health.

This chapter also presented an overview of the recent research on the use of ferrochrome slag for various engineering applications such as insulation bricks, refractories, road construction, cement/concrete, etc.

The production of industrial by-products is expected to increase rapidly in the coming few years due to the industrial development in countries around the world, which will have a detrimental impact on the environment unless waste management policies and legislations are developed. Therefore, successful utilization of industrial by-products in the construction industry will require more detailed economic and environmental analyses as well as engineering design specifications.

Acknowledgments

The authors would like to thank Prof. Munesh Chandra Adhikary, PG Council Chairman, Fokir Mohan University, for his invaluable guidance, advice, and constant inspiration throughout the entire process. They also would like to thank Mr. Mukteswar Mohapatra of Fokir Mohan University for his support. The authors convey their sincere thanks to MSCB University, Baripada, Odisha, India for encouraging the research work.

References

1. https://www.conserve-energy-future.com/causes-effects-of-industrial-pollution.php.
2. A Siddiqua, J.N. Hahladakis, Al-Attiya WAKA, (2022), An Overview of the Environmental Pollution and Health Effects Associated with Waste Landfilling and Open Dumping, Environmental Science Pollution Research International Vol. 29, pp. 58514-58536 (doi: 10.1007/s11356-022-21578-z).
3. C. K. Purchase, Al D. M. Zulayq, B. T. O'Brien, M. J. Kowalewski, A. Berenjian, A. H. Tarighaleslami, and M. Seifan, (2021) Circular Economy of Construction and Demolition Waste: A Literature Review on Lessons,

Challenges, and Benefits, Materials (Basel), Vol. 23, pp.76 –xxx (doi: 10.3390/ ma15010076).
4. https://cpcb.nic.in/openpdffile.php?id=TGF0ZXN0RmlsZS8xNTlfMTQ5NTQ0NjM5N19tZWRpYXBob3RvMTkyLnBkZg
5. U. Neethu and MD D. Anwer, (2018), Using of Ferro-Chrome Slag as a Coarse Aggregate and Stone Dust as a Fine Aggregate, International Research Journal of Engineering and Technology (IRJET), Vol. 05, pp. 686-692.
6. https://www.safefoodfactory.com/en/editorials/46-alloying-elements-stainless-steel-and-other-chromium-containing-alloys/
7. https://www.ispatguru.com/chromium-in-steels/
8. J. Gao, Y. Feng, D. Feng, Z. Zhang, and X. Zhang, (2020), Solidification with Crystallization Behavior of Molten Blast Furnace Slag Particle During the Cooling Process, International Journal of Heat and Mass Transfer, Vol. 146(5), pp. 118888-xxx (DOI: 10.1016/j.ijheatmasstransfer.2019.118888).
9. M. Panigrahi, S. Mohanty, R. I. Ganguly, R. R. Dash, (2018), An Advanced Cured High Carbon Ferrochrome Slag (HCFCS) Geopolymer (GP): A Constructional Materials, IOP Conference Series: Materials Science and Engineering, Vol. 410, pp. 012002-xxx.
10. C. Ugwuegbu, (2012), Technology Innovations in the Smelting of Chromite Ore, Innovative Systems Design and Engineering, Vol 3, pp. 48-54.
11. V. C. Keesara, (2009). Chromite. Retrieved from http://www.scribd.com/doc/17277588/Chromite-Lion Ferrochromium Field Trip Presentation, South Africa
12. D. L. Buchanan, (2001). Geological Setting of Nickel, Chromium and Vanadium Ore Deposits. In: Dunton, S. N. and Jones, M. J. (Eds). The Institution of Mining and Metallurgy.
13. O. Naiker, (2006). The Development and Advantages of Xstrata's Premus Process. Retrieved from http://www.pyrometallurgy.co.za/InfaconXI/011.pdf
14. The Full Wiki (2012). Electri Arc Furnace: Wikis. Retrieved from http://www.thefullwiki.org/Electric_arc_furnace
15. V. O. Rao, and B. K. Singh, (1996). Turn Around Story at Ferro Alloy Plant. Bamnipal, Tata Tech., Vol. 23, pp. 83-87.
16. R. P. Goel, (1997), Smelting Technologies for Ferrochromium Production – Recent Trends. NML, Jameshedpur –831 007, pp. 37-50.
17. J. Daavittila, M. Honkaniemi, and P. Jokinen, (2004). The Transformation of Ferrochromium Smelting Technologies during the Last Decades,SAIMM Proceedings of the Tenth International Ferroalloys Congress, pp. 541-549.
18. Xstrata (2012). Lion Ferrochrome Smelter. Retrieved from http://www.xstrataalloys.com/ EN/Operations/ Pages/LionFerrochromeSmelter.aspx
19. T. R. Curr, (2009). The History of DC Arc Furnace Process Development. Retrieved from http://www.mintek.co.za/Mintek75/Proceedings/C01-Curr.pdf
20. https://en.wikipedia.org/wiki/Ferrochrome

21. https://www.chemistrylearner.com/ferrochrome.html#:~:text=Ferrochrome%20Chemical%20and%20Physical%20Properties,The%20physical%20and&text=Atomic%20Mass%3A%20The%20atomic%20mass,alloy%20is%202642%20%C2%B0C.
22. "ASTM A482".
23. "Ferrochromium -- Specification and conditions of delivery - ISO 5448:1981"
24. Y. L. Holappa and Y. Xiao, Slags in ferroalloys production—review of present knowledge, The Journal of The South African Institute of Mining and Metallurgy, 2004, pp. 429-438 (The South African Institute of Mining and Metallurgy, 2004. SA ISSN 0038–223X/3.00 + 0.00. The paper was presented at the SAIMM Conference: VII International Conference on Molten Slags Fluxes and Salts, 25–28 January 2004).
25. https://www.ispatguru.com/ferro-chrome/
26. http://www.metallicalloys.in/ferro-chrome.html
27. https://www.ferro-alloys.com/en/News/Details/177287
28. P. J. Bhonde, A. M. Ghodgaonkar and R. D. Angal, Various Techniques To Produce Low Carbon Ferrochrome INFACON XI , pp. 85-90 (https://www.pyrometallurgy.co.za/InfaconXI/085-Bhonde.pdf).
29. https://www.wanhuametal.com/news/low-carbon-ferrochrome.html
30. http://etimet.com/en/ultra-low-carbon-ferrochrome-lc-fecr#:
31. M. G. Fontana, (1978), Corrosion Engineering", 2nd edition, Mc-Graw Hill International Book Company, New York and London, p 59, 60.
32. V.P. Elyutin, (1957), Production of Ferroalloys and Electrometallurgy", 2nd edition, The State Scientific and Technical Publishing House, Moscow, pp.190-196.
33. R.P. Ebraham, (1985), Extraction of Non-ferrous Metals, 1st edition, East West Press, New Delhi, pp. 216.
34. F. P. Edneral, (1979), Electrometallurgy of Steel and Ferroalloys, Vol. 2, 1st edition, Mir publishers, Moscow, pp. 151-152.
35. S. Shimizu, (1973), Low Carbon Ferrochrome, Japan Steel and Tube Corporation, (1971), 31, pp. 656 (Japanese), 'Chemical Abstract', Vol.79, pp. 8568.
36. G. B. Alcock, (1979), Metallurgical Thermochemistry, 5th edition, Maxwell, M.C. pp. 375, 379.
37. Chu, W. F.: "The Kinetics of the Reduction of Chromium Oxide by Hydrogen", Met. Trans. (B), (1979), Vol. 10 B, pp. 401-407.
38. Rankin W.J.: "The Kinetics of the Reduction of Chromite by Graphite and CO", Inst. of Min & Met. Transactions, C (1979), Vol.8 C, June, pp C 107.
39. N.S. Sundermurti, (1980), Kinetic of Chromite Ore Reduction with Carbon, Trans. of Ind. Inst. of Metals, Vol. 38, No. 5, October, pp. 423-425.
40. G. Hiroshi, (1977), Reduction of Chromite ($FeCr_2O_4$) with Carbon, Nippon Kinzoku Gakkaishi, Vol. 41, (3), pp. 275 (Japanese), Chemical Abstract, Vol. 86, pp. 158815 C.

41. V. Sheshadri, (1982), Kinetics of Reduction of Synthetic Chromite with Carbon, Tran. Iron and Steel Inst. Japan, Vol. 82, pp. 925-xxx.
42. J. C. Pahlman, Direct Method to Prepare Low Carbon Ferrochrome, Physical Chemistry of Extraction Metallurgy, Proc. Conference, New York, USA, 24-28, Feb. 1985, 'Metal Abstract', (1985), 06-42-0071.
43. S.K. Khalafalla, Direct Method to Prepare Low Carbon Ferrochrome, Proceedings of the conference of Physical Chemistry of Extraction Metallurgy, Publication of the Metallurgical Society of AIME, Pennsylvania (1985), pp. 433.
44. J. F. Elliot, (1976), Kinetics of Reduction of Cr_2O_3 by liquid Fe-Cr-C Alloys, Iron making Steel making, No. 5, pp. 268-xxx.
45. S. Ghose, J. K. Nanda, and B. B. Pater, Duplex process for production of low carbon ferrochrome, I. D. C. of Orissa Limited. Bhubaneswar, pp. 215-217.
46. M. Panigrahi, S. Mohanty, R. I. Ganguly, R. R. Dash "Synthesis And Characterizations Of High Carbon Ferrochrome (HCFC) Slag Based Geopolymer, IntechOpen Publisher, UK. ISBN: 978-1-83968-989-5 (DOI: http://dx.doi.org/10.5772/intechopen.97140).
47. M. Karhu, B. Talling, P. Piotrowska, *et al.*, (2020), Ferrochrome Slag Feasibility as a Raw Material in Refractories: Evaluation of Thermo-physical and High Temperature Mechanical Properties. Waste Biomass Val. 11, pp. 7147–7157 (https://doi.org/10.1007/s12649-020-01092-4).
48. https://www.indiamart.com/proddetail/ferro-chrome-slag-15803400548.html
49. Ferro chrome Manufacturers, Ferro chrome Exporters Delhi, India
50. P. Niemelä and M. Kauppi, production, characteristics and use of Ferrochromium slags, INFACON XI, pp. 171-179.
51. A. Rakhimov, (1965), Hydrochemical processing of helenite-bearing blast furnace slags into alumina, Thesis of Candidate Degree, Alma-Ata.
52. M. Almagambetov, S. Baisanov, D. Izbembetov, (2008), Improvement of low-carbon ferrochrome production technique, Promyshlennoct Kazakhstana, Vol. 46, p. 86-xxx.
53. A. Akuov, M. Tolumbekov, and D. Izbembetov, (2008), about the thermodynamics of metallothermy of ferrochrome, Transactions of International conference "Nauka I obrazovanie – vedushij factor strategii Kazakhstan 2030", Karaganda, pp. 94-xxx.
54. E. Jak and P. C. Hayes, (2008), Phase Equilibria Determination in Complex Slag Systems, Mineral Processing and Extractive Metallurgy IMM Transactions section C, Vol. 117, pp. xxx (DOI: 10.1179/174328508X272344).
55. A. Hayhurst, Crystallization Processes in a High-carbon Ferrochromium Slag, Vereeniging Refractories, South Africa, pp. 187-193.
56. S. A. Nightingale and B. J. Monaghan, (2008), Kinetics of Spinel Formation and Growth during Dissolution of MgO in $CaO\text{-}Al_2O_3\text{-}SiO_2$ Slag, Metallurgical and Materials Transactions B, Vol. 39B, pp. 643-xxx, (https://ro.uow.edu.au/engpapers/500).

57. L. Holappa and Y. Xiao, (2004), Slags in ferroalloys production-review of present knowledge, The Journal of The South African Institute of Mining and Metallurgy, pp. 429-438.
58. KC Mills: Fundamentals of Metallurgy, edited S Seetharaman, publ. Abington UK (200)
59. http://www.steelconstruction.info/images/8/89/BOS_diagram_cropped.PNG
60. K Mukai: Interfacial phenomena in Materials processing edit ED Hondros, M McLean and KC Mills, Book 692 Royal Society, publ IOM , London (1998) p. 201.
61. A. A. Kazakov, Y. Matveev, L. Arykova, and V. Ryabov Russian Metallurgy, (1993) (4) 94-96 see also A. A. Kazakov, (1997), Russian Metallurgy, Vol. 6, pp. 25-29.
62. S Riaz and KC Mills – REF?.
63. D.W. Bruce and N. S. Hunter: Proc. 2nd Europ. Conf. on Continuous Casting. Dusseldorf , (1994) pp. 156-161.
64. K. C Mills, Y. Su, A. B. Fox, Z. Li, R. P. Thackray, and H. Tsai, (2005), A review of slag splashing, ISI Journal of International. Vol. 45, pp. 619-633.
65. S. Oxley, P. N. Quested and K. C. Mills: The benefits of using experimental thermophysical property data on defect predictions Proc. AVS Conf. held Santa Fe , NM, Feb. (1999).
66. B. O. Mysen, (1988), Structure and Properties of silicate melts, publ Elsevier, Amsterdam.
67. B. O. Mysen, (1990) Earth Science Reviews, Vol. 27, pp. 281-365.
68. M. E. Fleet, X-ray diffraction and spectroscopic studies on the structure of silicate glasses, in short course on silicate melts edit. C. M. Scarfe, Mineralogical Soc. Ottawa, Canada, Oct (1986), pp.1-35.
69. K. C. Mills, (1993), ISIJ International, Vol. 33, pp. 148-156.
70. K. Mills, (2011), the estimation of slag properties, Southern African Pyrometallurgy, pp. 1-56.
71. D. Sichen, J. Bygden and S. Seetharaman, (1994), Metallurgical Transcection B, Vol. 25B, pp. 519-525 and 589-595.
72. T. Tanaka, M. Nakamoto, and J. Lee, Proc. Metal Separation Technology held Copper Mountain, Junre (2004) edited R E Aune andM Kekkonen publ. Helsinki Univ. Technol., pp. 135-142.
73. P. J. Bray, R. V. Mulkern, and E. J. Holupka, (1988), Journal of Non-Crystalline Solids, Vol. 75, pp. 29-36 and 37-44.
74. J. Zhongtang and T. Yongxing, (1992), Journal of Non-Crystalline. Solids, Vol. 146, pp. 57-62.
75. Y. Waseda and J. M. Toguri, (1998), the structure and properties of oxide melts, publ. World Scientific, Singapore.
76. Y. Sasaki, M Iguchi, and M Hino, (2007), ISIJ International,. Vol. 47, pp. 346-347 and 638-642 and 1370-1371.

77. T. Asada, Y. Yamada, and K. Ito, (2008), ISIJ International., Vol. 48, pp. 120-122.
78. G. H. Zhang and K. C. Chou, (2010), Metallurgical and Materials Transactions, Vol. 41B, pp. 131-136.
79. G. H. Zhang, B. J. Yan, K. C. Chou, and F. S. Li, (2011), Relation between viscosity and electrical conductivity of silicate, Metallurgical and Materials Transactions B
80. M. Barati and K. S. Coley, (2006), Metallurgical and Materials Transactions, Vol. 37B, pp. 41-49.
81. RJ Hundermark, S Jahanshahi and S Sun: Proc. IMPC held Cape Town South Africa 2003 and R. J. Hundermark Masters Thesis, Univ. of Cape Town (2003) cited in ref 98.
82. S. Jahanshahi, S. Sun, and L. Zhang: Proc.10th Intl. Ferroalloy Congress held Feb (2004) publ. SAIMM www.co.za./Journal/v104no9p529.pdf.
83. Slag Atlas, publ VDEh , Dusseldorf (1995).
84. Q. Jiao and N. J. Themelis, (1988), Metallurgical and Materials Transaction B, Vol. 19B, pp. 133-140.
85. J. Zhang, S. Shu and S. Wei: Proc.of the 6th Intl. Conf. Molten Slags, Fluxes and Salts held Stockholm June (2000).
86. Z. Y. Qiao, L. T. Kang, Z. M. Cao, and B. Y. Yuan, (2003), CAMP- ISIJ, Vol. 16, pp. 868-871.
87. M. Hayashi, H. Ishii, M. Susa, H. Fukuyama and K. Nagata, (2001), Physics and Chemistry of Glasses, Vol. 42, pp. 6-xxx.
88. K. Nagata and K. S. Goto, 2nd Intl. Symp.Metallurgical slags and fluxes edited HA Fine and D Gaskell held Lake Tahoe, NV publ TMS-AIME Warendale, PA (1984) pp. 875-889.
89. S. Ozawa and M. Susa, (2005), Iron making and Steel making, Vol. 32, pp. 487-493.
90. M Susa, M. atanabe, S Ozawa and R Endo, (2007), Iron making and Steel making, Vol. 34, pp. 124-130.
91. Y Kang and K Morita, (2006), ISIJ International, Vol. 46 pp. 420-426.
92. Y Kang and K Morita, private communication 2003.
93. M Susa, K Kubota, M Hayashi and K. C. Mills: (2001) Iron making and Steel making, Vol. 28, pp. 390-395.
95. S Ozawa, R Endo and M Susa, (2007), Tetsu-to-Hagane, Vol. 93, pp. 8-14.
96. K. Nishioka, T. Maeda, and M. Shimizu: (2006), ISIJ International Vol. 46, pp. 427-433.
97. C. H. Ajay and K. Durga Rani, (2022), Ferrochrome Slag characterization and its compatibility with Quarry dust in Flexible pavement, IOP Conf. Series: Earth and Environmental Science, Vol. 982, pp. 012008 (doi:10.1088/1755-1315/982/1/012008).
98. IS: 2386 Part (I, III, IV, V)-1963, Indian Standard Methods of Test for Aggregates for Concrete, Bureau of Indian Standards.

99. IS: 2386(part I)-1963(reaffirmed 1997), Methods of Test for Aggregates for concrete, Part 1- particle size and shape, Bureau of Indian Standard, New Delhi, India.
100. https://theconstructor.org/practical-guide/grain-size-analysis/2261/
101. Willem A Gericke, Development Manager, Chrome Alloys, Samancor Ltd, INFACON 7, Trondheim, Norway, June 1995, Eds.: Tuset, Tveit and Page-131-140, Publishers: FFF, Trondheim, Norway
102. W.A. Gericke, Promoting chromium is also defending chromium. Paper presented at the ICDA Spring Meeting. Paris. May 1994. (Unpublished).
103. Toxicological Profile for Chromium. TP-92/08. U.S. Department of Health & Human Services. pg 5. April 1993.
104. Idem pg 42.
105. Genmin Laboratories. Certificate of Analysis. November 1993. (Unpublished).
106. Health, Safety & Environmental Considerations for the Chromium Industry. International Chromium Development Association. pg 4. June 1994.
107. Ground water pollution investigation. Steffen, Robertson & Kirsten. Report 20023511. June 1994. (Unpublished).
108. Development of a risk-based environmental remediation programme for Ferrometals, Witbank. Steffen, Robertson & Kirsten. Report 20023512. June 1994. (Unpublished).
109. J. L. Broadhurst, Treatment of Cr (VI) - bearing wastes from ferrochromium plants: a theoretical evaluation. Gencor Process Research report 92/422. September 1992.
110. J. L. Broadhurst, Detailed evaluation of results in report R05/93. Gencor Process Research report no. 93/402. May 1993.
111. E. J. Solomon, et al., Leaching Test Work On Current Arisings for the Chrome Six task group. Gencor Laboratories report no R05/93-MF 03. May 1993.
112. E.J. Solomon, et al. Investigation into Analysis of Cr (VI) in the Samancor Chrome Alloys Division. Gencor Laboratories report no RlO/93 - SM 10. May 1993.
113. W.A. Apel. and C.E. Turick, (1991), Bioremediation of Hexavalent Chromium by Bacterial Reduction. Mineral Bioprocessing. The minerals, Metals & Materials Society. pp 376.
114. Conner, J.R. Stabilizing Hazardous Waste. Chemtech. pp. 35. December 1993.
115. Van Tienhoven, A.M. Study of the feasibility of the removal of chromium from soil by higher plants. University of the Witwatersrand. 1993 (Unpublished).

2

Production and Test Methods of Waste-Based Ceramics

Muktikanta Panigrahi[1]*, Ratan Indu Ganguly[2] and Radha Raman Dash[3]

[1]Department of Materials Science, Maharaja Sriram Chandra Bhanja Deo University, Keonjhar Campus, Odisha, India
[2]Department of Metallurgical Engineering, National Institute of Technology, Raurkela, Odisha, India
[3]National Metallurgical Laboratory (NML), Jamshedpur, Jharkhand, India

Abstract

In this chapter, a literature survey is presented on the production of ceramic and its products using waste materials such as silica fumes, low- and high-carbon fly ash, red mud, high- and low-carbon blast furnace slag, rice husk ash, zeolites, argon oxygen decarburization (AOD) slag, bottom ash, pond ash, etc. Also discussed are the different production methods adopted to prepare ceramics, including curing, sol-gel, deep mixing, hydrothermal synthesis and low-temperature calcination methods, mechanical activation (MA), blended mineral admixtures, an alkaline solution of thermally activated silica-alumina-bearing mineral with additives such as calcium carbonate, microwave radiation, and the sintering process. Plenty of data are cited to show the effect of variables on the production of ceramics.

Powder mixing is an important operation which is followed routinely by industries and pharmaceuticals. The quality of products depends on certain operating conditions such as the equipment, technical parameters and formulation. In this work, an operational qualification for powder mixing during formulation is specified. During dry powder mixing, the operation is aimed at optimizing mixing time for obtaining a homogeneous mixture of ceramic powder.

The powder compaction process of ceramic powders is carried out at room and elevated temperatures using a uniaxial die compaction machine. The parameters used are the force and displacement during their movement. This provides information on tooling movement and required force for compaction of powders. The density is obtained by measuring dimensions of green compact powder.

**Corresponding author*: muktikanta2@gmail.com

Muktikanta Panigrahi, Ratan Indu Ganguly and Radha Raman Dash (eds.) High Electrical Resistant Materials: Ferrochrome Slag Resource Ceramics, (63–82) © 2024 Scrivener Publishing LLC

A sintering method is used as part of the manufacturing system. A wide range of ceramic materials can be produced by this process. Sintered products affect the performance of the resultant component. To obtain best quality output, knowledge of the effects of sintering and post-processing methods is required.

Keywords: Industrial waste, fly ash, pond ash, slag, ceramics, synthesis, characterization techniques

2.1 Introduction

Ferrochrome slag is a waste material generated during the production of FeCr products. Production methods and sources of raw materials influence properties of slag material. Therefore, slag material has a complex structure. Hazardous gases are evolved during production of FeCr products. This causes environmental problems. Hence, utilization of slag is an important challenge in the area of solid waste management [1–3]. Ceramic production from slag is one of the effective ways to utilize it.

Conventional ceramic material is mainly prepared from natural minerals such as clays, feldspar, quartz, etc. Novel ceramics are prepared from industrial waste which has replaced natural minerals fully or partly. Traditional ceramics are made up from a ternary system (i.e., K_2O/Na_2O-SiO_2-Al_2O_3). Slag ceramics consist of multicomponent systems, i.e., CaO-MgO-Al_2O_3-SiO_2, CaOMgO-Al_2O_3-SiO_2-Fe_2O_3, and SiO_2-MgO-CaO-Fe_2O_3. Flexible ceramics consisting of CaO, MgO, and Fe_2O_3 adjust the pore structure of slag ceramics (above 10%). Solid waste (one or more as a raw material) with proper proportions is utilized to make ceramic having good sound absorption, thermal insulation, formation of crystalline phases, etc. [4–8].

Some slag ceramics may have limited uses because of their high CaO content, which may limit the firing process. High CaO content slag suffers due to low melting temperature [4, 9]. However, MgO-bearing slag increases sintering temperature (>1200 °C), which increases production cost [10]. Due to high sintering operation, volatile components bubble out to create pinholes on the surface and inside ceramic material. This in turn affects the performance of ceramic material.

The adopted ceramic production techniques, such as curing [11], sol-gel [12], deep mixing [13], hydrothermal synthesis [14] and low-temperature calcination methods [15], mechanical activation (MA) [16], blended mineral admixtures [17], an alkaline solution of thermally activated silica-alumina-bearing mineral with additives such as calcium carbonate [18], innovative *in-situ* co-reticulation process [19], an innovative method to reclaim waste molding sands containing water glass with "dry" or "wet" activation of inorganic binder in waste molding sand mixtures physically hardened by microwave radiation [20], sintering process [21], etc., are adopted to prepare value-added ceramic products.

Solid wastes for production of value-added products are silica fumes [22], low- and high- carbon fly ash [23, 24], red mud [25], low- and high-carbon blast furnace slag [26, 27], rice husk ash [28], zeolites [29], argon oxygen decarburization (AOD) slag [30], bottom ash [31], pond ash [32], man-made rock [33], biochar concrete [34], attapulgite pozzolana [35], copper mine tailings [36], activated blast furnace slag [37], low Ca electric arc ferronickel slags [38], etc.

Reviews are available on the preparation of slag ceramics on a laboratory scale. This is a primary stage. Actual production of slag ceramics will come to the forefront in the near future. Some reports on slag ceramics are cited below.

Duran Suarez [39] prepared new ceramic materials using tungsten mining wastes (region of Beira Interior in Portugal). New ceramic products are prepared trough sintering process in the temperature range of 800 °C to 1300 °C. Mineralogical characterizations, morphologic evaluation, and mechanical testing were carried out on new products.

Singh *et al.* [40] studied room-temperature setting phosphate-bonded ceramics. Phosphates of Mg, Mg-Na, Al and Zr were studied to stabilize ash surrogate waste containing RCRA metals (as nitrates and RCRA organics). Different tests were performed to characterize and evaluate performance.

Arenas *et al.* [41] developed a material composed of 80% w/w ceramic wastes. The developed material was applied in the manufacture of road traffic noise reducing devices. Characterization of acoustic, physical, and mechanical properties was performed.

Boccaccini *et al.* [42] manufactured useful glass-ceramic products from industrial waste materials. Controlled crystallization supports improvement properties of vitrified products. It has solved the burden of the increased amount of slag, fly ash and combustion dust generated by glass-ceramic unit.

Stolboushkin *et al.* [43] produced building ceramics using coal waste obtained from mining enterprises (Kemerovo region of Siberia). Results of building ceramics, such as the chemical and mineralogical compositions of waste coal, are discussed.

Mendes *et al.* [44] prepared basaltic waste-based ceramics. Various characteristic investigations were made to characterize products.

He *et al.* [45] highlighted steelmaking slag (SS) as an industrial waste. Glass ceramics are produced from SS. Such ceramics have good properties.

Trezza *et al.* [46] developed ceramic using demolition ceramic wastes (DCWs). The DCWs were investigated by different characterizations in order to determine their potential use as supplementary cementitious materials. Results indicated the possibility of using them as a filler material. As hydration progresses, the pozzolanic activity of ceramic waste contributes to the strength of properties.

Bateman and Solbrig [47] developed a process for preparing a nuclear container developed from two waste streams (stainless steel based and ceramic based). Performance of container made from ceramic waste was assessed, with encouraging results.

Back *et al.* [48] promoted recycling of industrial waste using steel industry slags and power plant waste. They developed value-added products from these resource waste materials. There is an opportunity to manufacture glass-ceramic materials of SiO_2-CaO-Al_2O_3 system by different heat treatment processes.

Loida and Schubert [49] developed aluminum-silicate-based ceramic materials and evaluated their performance.

Presotto and Mymrine. [50] developed new ceramic materials using serpentine, glass waste and clay red. The new ceramic that was developed has six formulations on serpentine and red clay. The ceramic material is formed by uniaxial pressing followed by firing in an electric oven at temperatures of 1100 °C, 1200 °C, 1250 °C and 1300 °C.

Biff and Silva [51] prepared ceramic tiles using Brazilian standard NBR 13818 (1997).

Oversby *et al.* [52] developed ceramic product by ceramic processing methods. Treatment process residues were mixed with ceramic precursor materials, dried, calcined, and formed into pellets at room temperature. It was then sintered at 1150–1200 °C to produce final products.

Harker and Flintoff [53] prepared polyphase ceramic and glass-ceramic from simulated Idaho Chemical Processing Plant wastes. They reduced sintering temperature by adding calcined waste and chemical additives. Major phases of ceramic were found to be CaF_2, $CaZrTi_2O_7$, $CaTiO_3$, monoclinic ZrO_2, and amorphous intergranular material. Glass-ceramic contains crystalline CaF_2, Al_2O_3, and $ZrSiO_4$ dispersed in the glass matrix. Natural mineral zircon is a stable host for tetravalent actinides.

Marangoni [54] focused on development of a wide range of ceramic components. The sinter-crystallization process produced glasses from waste using fly ash, slag, and sludge.

Toropkov *et al.* [55] discussed various methods of synthesizing hydroxyapatite-based ceramic materials. The compatibility and suitability of the material developed for implantation were investigated.

Jirajariyavej *et al.* [56] worked on the challenge of ceramic restoration faced by dentists. They evaluated the influence of abutment material and ceramic thickness on the final color of different ceramic systems.

Șoim *et al.* [57] assessed the translucency of two pressable ceramics and examined their masking property. Millimeter-sized thickness discs were fabricated using two pressable ceramics (shade/translucency): 1M1T/HT, 1M2T/HT, 2M2T (VITA PM9), and A1LT/HT, B1LT/HT, A2LT. High translucent ceramic is used in restoring the appearance of dental structures.

Diana *et al.* [58] fabricated ceramic membranes composed of fly ash and clay using low sintering temperatures (900 °C and 1000°C). Membranes were prepared by using different proportions of fly ash and clay, i.e., 35% and 65% (M1), 45% and 55% (M2), 50% and 50% (M3), 55% and 45% (M4), 65% and 35% (M5). Pore sizes, structures, and composition were studied by SEM, EDX and XRD.

Ma *et al.* [59] reviewed fly ash obtained from China's coal-fired power plants. They studied the utilization of fly ash in different areas such as valuable elements extraction, geopolymer production, fly ash-based ceramics synthesis and soil desertification control.

Maitra [60] studied the different production techniques emerging globally for production of value-added ceramics from fly ash, glassy materials, porcelains, refractories, etc.

Ercenka *et al.* [61] prepared glass and glass-ceramics from boron waste powder and fly ash in different proportions. Powder mixtures of glass and glass-ceramics were sintered at 1500 °C in an electric furnace. Sintering of the mixture transformed the material into glass-ceramic through crystallization process. Crystallization and glass-transition temperatures of materials were examined by DTA (differential thermal analysis). Annealing of the product was done at 800 °C, 900 °C, and 1000 °C for 1 h before performing XRD analyses. XRD analyses were carried out on the products. The predominant phases identified were diopside augite and tincalconite. Also, residual glassy phases were detected. Mechanical tests, such as hardness and fracture toughness, were done on the products. Microstructures of glass-ceramic materials were examined with SEM.

Carabba *et al.* [62] developed innovative and sustainable alkali-activated composites with enhanced performance at elevated temperatures. They indicated a reduction in production cost for their products. For composite preparation, they used coal fly ash as a precursor for alkali activation. Also, they used recycled refractory particles to enhance thermal stability of products. Quantitative mineralogical analyses were performed on products. Phases were identified before and after thermal exposure. Results indicated better dimensional stability of composites up to 1240 °C.

Yang *et al.* [63] produced cost-effective porous ceramics using fly ash as a main ingredient and alumina as pore-forming agent. Different parameters like line shrink, bulk density, mechanical strength, porosity, and phase composition were determined. Properties obtained were due to the addition of alumina, and were also obtained due to formation of mullite (alumina can react with SiO_2).

2.2 Ceramic Process Parameters

Process parameters will govern performances of ceramic products. Therefore, they play a critical role in the production of quality ceramic materials. Process parameters are classified into three major groups: parameters during mixing, parameters during compaction, and parameters during sintering. Each category is described below.

2.2.1 Parameters During Mixing

Quality of end-products depends on degree of mixing of constituent materials [64, 65]. Uniform mixing plays a key role in obtaining quality products [66, 67]. Principally, physical features of powders, such as purity, particle density, shape, size/ratio, surface properties and intensity of cohesion, play a vital role in controlling mixture segregation. Optimization of composition (i.e., qualitative and quantitative) of the design is well supported to prepare on a pilot scale production. In relation to optimization of constituents, production process parameters include techniques and traceability requirements. Operated equipment and formulation (such as mixer design) and equipment operation in different unit operations are also important for getting quality product [68-70]. During unit operations, variables such as load, speed and mixing time are important, since blending uniformity is the main objective [71]. Many reports have discussed the technologies for determining mixture homogeneity and optimal blending time. Two of the options are in-line (by real-time monitoring of mixing) or off-line (by using strategies) to take periodic samples from the mixers [72, 73].

2.2.2 Parameters During Compaction

The powder compaction process is a process of forming powder material to a desired shaped material through an applied compacting force. The main objective of the compaction process is to achieve high-density green compact with required shape and dimension. Darren W. Smith [74] has described three stages of compaction:

Stage I: Bulk quantities of particles undergo packing more efficiently.
Stage 2: Elastic and plastic deformation is imparted to the particles.
Stage 3: Particles may fracture if compacting load is increased [75].

The mechanism of compaction or densification is that since the probability of additional bulk particle rearrangement is sufficiently low, substantial particle deformation may take place, resulting in further densification, which leads to the reduction of the porosity of the compacted mass. Factors that control the extent of plastic deformation are the ductility, yield strength and strain hardening characteristics of the constituent particles.

In addition, warm compaction is a type of compaction process in which a temperature gain increases the densification during compaction. During the warm pressing, powder has to be mixed with binder in order to get good interaction with constituent particles. Warm compaction technique is a process of pressing a preheated powder in a heated die. Warm-pressed powders can achieve a density 0.1–0.2 g/cm^3 or higher if compared with similar materials compacted at normal condition [76]. Compacted powder will undergo a volume expansion. It is very important to determine green density of compacted mass [77]. By increasing temperature, volume expansion of the compacted powder is significantly reduced. Reduction of volume expansion is due to reduction of the springback parallel to the compaction axis, which results in the increase in green density of compacted powder. Cylindrical form/rectangular form of compacted mass are prepared in the laboratory. This is a single action die compaction, in which pressure is applied hydraulically.

2.2.3 Parameters During Sintering

In the powder metallurgy process, industrial piece(s) are produced through compacting powder form and sintering within a controlled setting. The high pressure required, sintering temperature, i.e., below the melting point, and sintering time affect the properties and processing cost. In typical sintering temperature, the driving force is reduced and the surface energy of the particles is caused by decreasing their vapor-solid interfaces. In the diffusion process, pores in the green compact mass diminish or even close up, resulting in the densification of the part, improving its mechanical properties, i.e., smaller powders and greater product strength are achieved

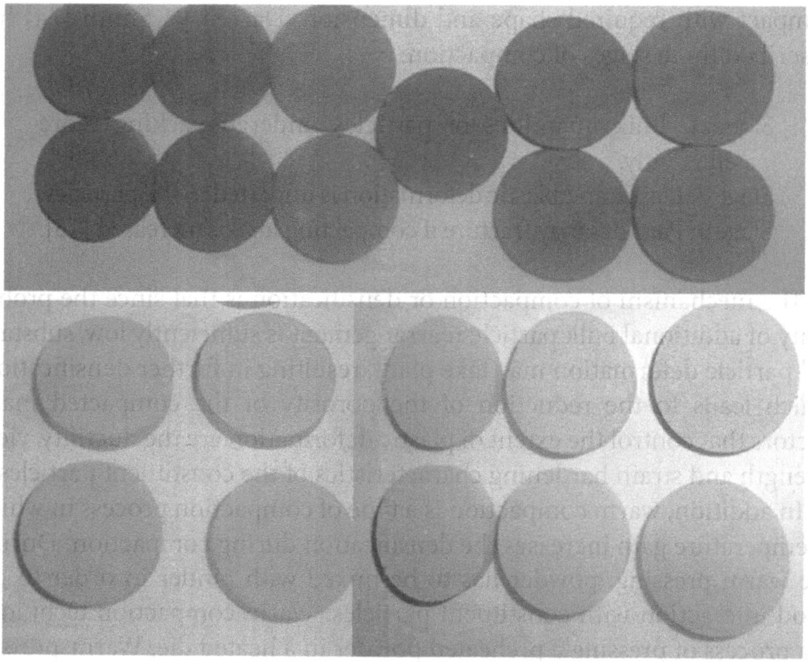

Figure 2.1 Sintered products (prepared from our laboratory) using a muffle furnace.

[78–81]. Parameters such as sintering temperature, pressure, time, powder density, and bit size are important to achieve good sintered mass. Sintering (firing) of pure oxide ceramics requires a relatively long time and high temperature because the diffusion proceeds in solid state. Applying pressure decreases sintering time and the resulted porosity.

Sintering process is conducted in different atmospheres (i.e., air, inert atmosphere) in a furnace [82]. Figure 2.1 shows some sintered samples prepared in the laboratory.

2.3 Characterization Techniques

Sintered materials characteristics are evaluated using XRD, FTIR, optical, SEM/EDS, TEM, mechanical measurement (such as Vickers hardness) and two-probe electrical resistivity determination techniques. The techniques are subsequently described.

2.3.1 X-Ray Diffraction (XRD) Technique

X-Ray diffraction is an important characterization technique for R&D studies [83]. It assesses parameters such as crystal structure, density, interplanar spacing, indices of planes (hkl), unit cell angle, unit cell dimension (a, b, c), etc.

Identification of the above parameters are made using Bragg's law (i.e., $n\lambda = 2d\sin\theta$). Both qualitative and quantitative analyses are possible using XRD technique. The grain size or particle size are determined using intensity spectra and line-breath at half intensity maxima. Thus, it is an indispensable technique for fundamental studies or analyses or phases presenting materials. Quantitative assessment is also made in non-crystalline phase, i.e., glassy phase through line profile.

X-rays are generated from CuKα target. X-ray energy is estimated with the expression [84, 85]

$$E = h\upsilon = h\frac{c}{\lambda} \tag{2.1}$$

Here,

h is Planck's constant (6.62×10^{-34} joule)
c is velocity of light (3×10^{8} m/s)
E is energy of the radiation

The X-ray wavelength is comparable to the size of atoms.

X-ray diffraction is mainly based on constructive interference of monochromatic X-rays and prepared samples. Incident X-rays and sample interaction produces constructive interference. The diffracted rays that are produced obey Bragg's law. Sample is scanned through a range of 2θ angles. All possible diffraction directions of lattice should be attained due to random orientation of powdered sample. Conversion of diffraction peaks to d-spacing allows identification of materials because of a set of unique d-spacing with standard reference patterns.

The present work records X-ray diffraction patterns using CuKα radiation. Operation parameters (i.e., 40 kV and 20 mA) are kept fixed. Powder samples are put on a quartz sample holder at room temperature and are scanned at diffraction angle 2θ from 10° to 70°.

2.3.2 Fourier Transform Infrared (FTIR) Spectroscopy

Fourier transform infrared (FTIR) spectroscopy is also another non-destructive technique (NDT). It is used to identify chemical groups in the test sample. An FTIR study is carried out using an FTIR spectrometer (Nicolet Nexus 870, Thermo Fisher Scientific Inc., USA). The FTIR spectrometer parameters (50 scan at 4 cm^{-1} resolution, transmittance/absorbance mode) are fixed. The spectrometer [86] uses IR radiations (obtained from IR source) that are passed through the sample. Adsorbed IR energy is transmitted and recorded by a suitable detector. It is guided through an interferometer where a Fourier transform is performed on the output signal.

Before FTIR analyses, test powder sample is used. Pellet is prepared using a small amount of test sample and bulk amount of KBr. Both samples are ground with a mortar and pestle. The ground sample is placed in a die and then placed in the hydraulic pressure. The samples are compacted by applying 5 kg of pressure. The dimensions of such pellet is found to be 13 mm (diameter) and 0.3 mm (thick). Background is collected before running the sample. Prepared pellet is placed in a sample holder. Pellet is exposed to IR radiation (IR spectrometer) and data are collected.

2.3.3 Scanning Electron Microscopy (SEM)

Sample surface topography is studied with a SEM instrument which uses backscattered electrons in place of light to produce surface morphology [87]. During the study, an electron beam is generated from an electron gun that is focused on a small portion of the sample. Vacuum is maintained. The detector collects output signals. The signal is sent to a computer and displays the final image.

A special sample preparation technique is employed before SEM analysis. Non-conducting sample requires a thin layer of gold or platinum coating. Coating is performed using a sputter coater and inert environment is maintained using argon gas in the coating system. Test specimen(s) are placed in a coating chamber. Due to the application of electric field, argon gas is ionized and forms argon ion (Ar^+). Argon ions knock gold atoms from the surface of the gold foil; and these gold atoms settle onto the surface of the sample, producing a gold coating.

Chemical composition analysis is made with energy-dispersive X-ray spectroscopy (EDX) or/and wavelength-dispersive X-ray spectroscopy (WDXS) analysis [87]. EDX or WDXS is attached to SEM. Data are collected over a selected area of sample surface. Two-dimensional spectrum is generated.

2.3.4 Two-Probe Electrical Resistance Measurement (at Room Temperature)

In the present work, resistance of sintered sample is measured to show the electrical nature. For this measurement a two-probe setup is used, as schematically shown in Figure 2.2.

Generally, resistance of sample is measured by the two-probe method. Linear two-probe technique is used to measure resistance at room temperature. For this work, pellet sample is used to measure resistance. Two electrical contacts are made, which are done through copper wires and silver paste. Probe contact is made linearly with distance (S) from each other. Output voltage is proportional with applied current. Such contacts are known as ohmic contact. In this method, resistivity (ρ) is estimated by the following expression [88]:

$$\rho = \left(\frac{RA}{l} \right) \tag{2.2}$$

Here, S is probe distance in centimeter (*cm*), which is kept constant, *I* is supplied current in milliampere (*mA*) and corresponding voltage is measured in volt (*V*) or millivolt (*mV*).

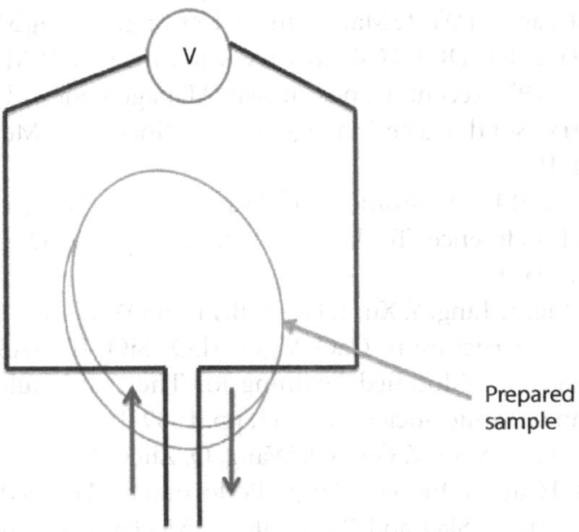

Figure 2.2 Schematic diagram of the linear two-probe setup for measuring electrical resistance.

2.4 Conclusion

General background information on ceramics was discussed in this chapter. The literature relevant to thermal power-plant-based ceramics and FeCr slag-based ceramics were presented in brief. Various raw materials and different production methods of ceramics were reviewed in detail. It is emphasized that there lies a major gap in research which deals with fruitful utilization of industrial waste and ferrochrome wastes. Therefore, a research goal is fixed to develop new ceramic products using thermal power plant waste and ferrochrome wastes.

Acknowledgments

The authors would like to thank Prof. Munesh Chandra Adhikary, PG Council Chairman, Fokir Mohan University, for his invaluable guidance, advice, and constant inspiration throughout the entire process. They also would like to thank Mr. Mukteswar Mohapatra of Fokir Mohan University for his support. The authors convey their sincere thanks to MSCB University, Baripada, Odisha, India for encouraging the research work.

References

1. E. Gunay, B. Kavaklı, B. Kavaklı, S. M. Tayfur, K. Eren, and S. A. Yıldırım, (2004), Steel Slag and Waste Management, Key Engineering Materials, pp. 264-268, pp. 2481-2484 (DOI: 10.4028/www.scientific.net/KEM.264-268.2481).
2. S. Kumar, (2019), Recent Trends in Slag Management & Utilization in the Steel Industry, Solid Waste Management, Minerals & Metals Review Vol. 2019, pp. 94-102.
3. A. Kumar, (2014), Utilization of Wastes from Integrated Steel Plant with Special Reference To India, Vol. 03, pp. 47-52 (DOI: 10.17577/IJERTV3IS10033).
4. K. Jiang, M. Xia, Y. Tang, Y. Xu, T. Deng, B. Li, and W. Chen, (2023), Formation of Closed Pore Structure in CaO-MgO-Al_2O_3-SiO_2 (CMAS) Porous Glass-Ceramics via Fe_2O_3 Modified Foaming for Thermal Insulation, Journal of the European Ceramic Society, Vol. 43, pp. 1689-1697.
5. M. Li, F. Zheng, Y. Xiao, Y. Guan, J. Wang, Q. Zhen, Y. Yu, (2021), Utilization of Residual Heat to Prepare High Performance Foamed Glass-Ceramic from Blast Furnace Slag and its Reinforce Mechanism, Process Safety and Environmental Protection, Vol. 156, pp. 391-404.

6. S. Shang, H. Fan, Y. Li, L. Li and Z. Li, (2022), Preparation of Lightweight Ceramsite from Solid Waste Using SiC as a Foaming Agent, Materials, Vol. 15, pp. 325-17 (https://doi.org/ 10.3390/ma15010325).
7. H. You, H. Sun, T. Peng, X. Zhou, L. Chao, and C. Wang, (2023), Effects of Illitic Clay on the Phases, Microstructure, Physical Properties and Pyroplastic Deformation of Industrial Slag Ceramics, Materials, Vol.16, pp. 233-17 (https://doi.org/ 10.3390/ma16010233).
8. R. He, N. Dai, and Z. Wang (2020), Thermal and Mechanical Properties of Geopolymers Exposed to High Temperature: A Literature Review, Vol. 2020, pp.1-17 (https://doi.org/10.1155/2020/7532703).
9. L. Zhao, Y. Li, Y. Zhou, D. Q. Cang, (2014), Preparation of Novel Ceramics with High CaO Content from Steel Slag, Materials and Design, Vol. 64, pp. 608-613 (DOI: 10.1016/j.matdes.2014.08.015)
10. H. An, F. Shen, X. Jiang, Y. Zhou, H. Zheng , and Q. Gao, (2022), Effects of Fine MgO-Bearing Flux on the Strength of Sinter before and after Low-Temperature Reduction, ACS Omega, Vol.7, pp. 8686–8696 (https://doi.org/10.1021/acsomega.1c06722).
11. E. M. M. Ewais, Y. M. Z. Ahmed, and A. M. M. Ameen, (2009), Preparation of Porous Cordierite Ceramic using A Silica Secondary Resource (Silica Fumes) for Dust Filtration Purposes, Journal of Ceramic Processing Research, Vol. 10, pp. 721-728.
12. T. Sathanandam, P. O. Awoyera, V. Vijayan, and K. Sathishkumar, (2017), Low carbon building: Experimental insight on the use of fly ash and glass fibre for making geopolymer concrete, Sustainable Environment Research, Vol. 27, pp. 146-153 (https://doi.org/10.1016/j.serj.2017.03.005).
13. B. Cetin, A. H. Aydilek, and Y. Guney, (2010), Stabilization of Recycled base Materials with High Carbon Fly Ash, Resources Conservation and Recycling, Vol. 54, pp. 878-892 (DOI: 10.1016/j.resconrec.2010.01.007).
14. H. Sun, C. Chen, L. Ling, S. Ali Memon, Z. Ding, W. Li , L.-P. Tang, and F. Xing, (2019), Synthesis and Properties of Red Mud-Based Nanoferrite Clinker, Journal of Nanomaterials, Vol. 2019, pp. 1-12 (https://doi.org/10.1155/2019/3617050).
15. N. You, J. Shi, and Y. Zhang, (2020), Corrosion Behaviour of Low-Carbon Steel Reinforcement in Alkali-Activated Slag-Steel Slag and Portland Cement-Based Mortars Under Simulated Marine Environment, Corrosion Science, Vol. 175, pp. 108874-xxx (https://doi.org/10.1016/j.corsci.2020.108874).
16. I. Narasimha Murthy, N. Arun Babu, and J. Babu Rao, (2016), High Carbon Ferro Chrome Slag – Alternative Mould Material for Foundry Industry, Procedia Environmental Sciences, Vol. 35, pp. 597-609 (https://doi.org/10.1016/j.proenv.2016.07.046).
17. L. Sun and K. Gong, (2001), Silicon-Based Materials from Rice Husks and Their Applications, Industrial & Engineering Chemistry Research, Vol. 40, pp. 5861–5877 (https://doi.org/10.1021/ie010284b).

18. M. Granda-Valdés, A. I. Pérez-Cordoves, and M. E. Díaz-García, (2006), Zeolites and Zeolite-based Materials in Analytical Chemistry, TrAC Trends in Analytical Chemistry, Vol. 25, pp. 24-30 (https://doi.org/10.1016/j.trac.2005.04.016).
19. E.-J. Moon and Y. C. Choi, (2019), Carbon Dioxide Fixation via Accelerated Carbonation of Cement-based Materials: Potential for Construction Materials Applications, Construction and Building Materials, Vol. 199, pp. 676-687 (https://doi.org/10.1016/j.conbuildmat.2018.12.078).
20. M. Cabrera, J. L. Díaz-López, F. Agrela, and J. Rosales, (2020), Eco-Efficient Cement-based Materials using Biomass Bottom Ash: A Review, Applied Science, Vol. 10, pp. 8026-24 (doi:10.3390/app10228026).
21. M. K. Panigrahi, P. K. Rana, A. K. Pradhan, P. K. Rout, A. K. Samal, S. Gupta and Mv B. Kumar, (2015), Production of Geopolymer based Construction Material from Pond Ash: An Industrial Waste, 19th International Conference Non-Ferrous Metal-2015 (9,10th July), pp. 190-200.
22. J. Davidovits, (1994), Geopolymers: Man-made Rock Geosynthesis and the Resulting Development of Very Early High Strength Cement, Geopolymer Cement/Concrete.
23. L. Chen, Y. Zhang, L. Wang, S. Ruan, J. Chen, H. Li, J. Yang, V. Mechtcherine, and D. C. W. Tsang, (2022), Biochar-Augmented Carbon-Negative Concrete, Chemical Engineering Journal, Vol. 431, pp. 133946 (https://doi.org/10.1016/j.cej.2021.133946).
20. T. Shi, Y. Liu, Y. Zhang, Y. Lan, Q. Zhao, Y. Zhao, and H. Wang, (2022), Calcined Attapulgite Clay as Supplementary Cementing Material: Thermal Treatment, Hydration Activity and Mechanical Properties, International Journal of Concrete Structures and Materials, Vol. 16, pp. 1-10 (https://doi.org/10.1186/s40069-022-00499-8).
24. R. S. Krishna, F. Shaikh, J. Mishra, G. Lazorenko, and A. Kasprzhitskii, (2021), Mine Tailings-based Geopolymers: Properties, Applications and Industrial Prospects, Ceramics International, Vol. 47, pp. 17826-17843 (https://doi.org/10.1016/j.ceramint.2021.03.180).
25. T. Luukkonen, Z. Abdollahnejad, J. Yliniemi, P. Kinnunen, and M. Illikainen, (2018), Comparison of Alkali and Silica Sources in One-Part Alkali-Activated Blast Furnace Slag Mortar, Journal of Cleaner Production, Vol. 187, pp. 171-179 (https://doi.org/10.1016/j.jclepro.2018.03.202).
26. K. Komnitsas, D. Zaharaki, and V. Perdikatsis, (2007), Geopolymerization of Low Calcium Ferronickel Slags, Journal of Materials Science, Vol. 42, pp. 3073-3082 (DOI: 10.1007/s10853-006-0529-2).
27. P. G.Asteris, P. B. Lourenço, P. C. Roussis, C. E. Adami, D. J. Armaghani, C. L. Cavaleri, E. Chalioris, M. M. Hajihassani, E. Lemonis, Ahmed S. Mohammed, and K. Pilakoutas, (2022), Revealing the Nature of Metakaolin-based Concrete Materials using Artificial Intelligence Techniques, Construction and Building Materials, Vol. 322, pp. 126500-xxx (https://doi.org/10.1016/j.conbuildmat.2022.126500).

28. Z. Chen, D. Li, W. Zhou, and L Wang, (2010), Curing Characteristics of Ceramic Stereolithography for an Aqueous-based Silica Suspension, Proceedings of the Institution of Mechanical Engineers Part B Journal of Engineering Manufacture, Vol. 224, pp. 641-651 (DOI: 10.1243/09544054JEM1751).
29. A. Nazeri, E. Bescher, and John D. Mackenzie, (1993), Ceramic Composites by the Sol-Gel Method: A Review, Ceramic Engineering & Science Proceedings, Vol. 14, pp. 1-19.
30. Sk S. Hossain and P.K. Roy, (2020), Sustainable Ceramics Derived from Solid Wastes: A Review, Journal of Asian Ceramic Societies, Vol. 8, pp. 984-1009 (https://doi.org/10.1080/21870764.2020.1815348)
31. J. Ortiz, C. Gomez-Yañez, R. López-Juárez, I. A. Velasco, and D. H. Pfeiffer (2012), Synthesis of Advanced Ceramics by Hydrothermal Crystallization and Modified Related Methods, Journal of Advanced Ceramics, Vol. 1, pp. 204-220 (DOI: 10.1007/s40145-012-0022-0)
32. P. Perumal, A. Hasnain, T. Luukkonen, P. Kinnunen, and M. Illikainen, (2021), Role of Surfactants on the Synthesis of Impure Kaolin-based Alkali-Activated, Low-Temperature Porous Ceramics, Open Ceramics, Vol. 6, pp. 100097-9 (https://doi.org/10.1016/j.oceram.2021.100097).
33. V. Berbenni, A. Marini, and G. Bruni, (2001), Effect of Mechanical Activation on the Preparation of $SrTiO_3$ and Sr_2TiO_4 Ceramics from the Solid State Systems $SrCO_3$-TiO_2, Journal of Alloys and Compounds, Vol. 329, pp. 230-238 (DOI: 10.1016/S0925-8388(01)01574-2).
34. M. Amin, B. A.Tayeh, and I. S. Agwa, (2020), Effect of using Mineral Admixtures and Ceramic Wastes as Coarse Aggregates on Properties of Ultrahigh-Performance Concrete, Journal of Cleaner Production, Vol. 273, 10 pp. 123073-xxx (https://doi.org/10.1016/j.jclepro.2020.123073).
35. T. Jiang, G. Li, G. Qiu, X. Fan, and Z. Huang, (2008), Thermal Activation and Alkali Dissolution of Silicon from Illite, Applied Clay Science, Vol. 40, pp. 81-89 (DOI:10.1016/j.clay.2007.08.002).
36. F. Colangelo, G. Roviello, L. Ricciotti, C. Ferone and R. Cioffi, (2013), Preparation and Characterization of New Geopolymer-Epoxy Resin Hybrid Mortars, Materials, Vol. 6, pp. 2989-3006 (DOI: 10.3390/ma6072989).
37. R. R. Mishra and A. K. Sharma, (2016), Review on Microwave-Material Interaction Phenomena: Heating Mechanisms, Challenges and Opportunities in Material Processing, Composites Part A: Applied Science and Manufacturing, Vol. 81, pp.78-97 (https://doi.org/10.1016/j.compositesa.2015.10.035).
38. S. V. Fomichev, N. P. Dergacheva, A. V. Steblevskii, and V. A. Krenev, (2011), Production of Ceramic Materials by the Sintering of Ground Basalt, Theoretical Foundations of Chemical Engineering, Vol. 45, pp. 526-529 (DOI: 10.1134/S0040579510051124).
39. J. A. Duran Suarez, (2014), Validation of New Ceramic Materials from Mining Waste Rocks. Mechanical Properties. Bulletin of the Spanish Society of Ceramics and Glass, Vol. 53, pp. 279-288 (http://hdl.handle.net/10481/34821).

40. D. Singh, A. S. Wagh, J. C. Cunnane, and J. L. Mayberry, (1994), Chemically Bonded Phosphate Ceramics for Low-Level Mixed Waste Stabilization, Proceedings of the American Chemical Society Symposium on Emerging Technologies in Hazardous Waste Management VI, held in Atlanta, September pp. 19- 21,
41. C. Arenas, L.F. Vilches, C. Leiva, B. Alonso-Fariñas, and M. Rodríguez-Galán, (2016), Recycling Ceramic Industry Wastes in Sound Absorbing Materials, Materiales de Construcción, Vol. 66, pp. e106-xxx.
42. A.R. Boccaccini, M. Köpf, and W. Stumpfe, (1995), Glass-Ceramics from Filter Dusts from Waste Incinerators, Ceramics International Vol. 21, pp. 231-235 (https://doi.org/10.1016/0272-8842(95)99787-C)
43. A. Y. Stolboushkin, D. V. Akst, O. A. Fomina, A. I. Ivanov, and V. A. Syromyasov, (2017), Analysis of waste coal from the enterprises of Kemerovo region as raw materials for production of ceramic materials, September, IOP Conference Series Earth and Environmental Science, Vol. 84, pp. 012037-xxx (DOI: 10.1088/1755-1315/84/1/012037).
44. T. M. Mendes, G. Morales and Paulo José dos Reis, (2016), Use of Basaltic Waste as Red Ceramic Raw Material, Ceramica, Vol. 62, pp. 157-162 (DOI: 10.1590/0366-69132016623621955).
45. F. He , Y. Fang, J. Xie, and J. Xie, (2012), Fabrication and Characterization of Glass–Ceramics Materials Developed from Steel Slag Waste, Materials & Design, Vol. 42, pp. 198-203 (https://doi.org/10.1016/j.matdes.2012.05.033).
46. M. A. Trezza, Silvina V. Zito, Alejandra Tironi, Edgardo Fabian Irassar, and V. Rahhal, (2017), Portland Blended Cements: Demolition Ceramic Waste Management, Materiales de Construcción, Vol. 67, pp. 114-xxx (DOI: 10.3989/ mc. 2017.00516)
47. K. J. Bateman and C. Solbrig, (2008), Stabilizing Glass Bonded Waste Forms Containing Fission Products Separated from Spent Nuclear Fuel, Separation Science and Technology, Vol. 43, pp. 2722-2746 (DOI: 10.1080/01496390802151831).
48. G.-S. Back, H.-S. Park, Seo, M. Sung, and W.-G. Jung, (2015), Exploring High-Strength Glass-Ceramic Materials for Up Cycling of Industrial Wastes, Metals and Materials International, Vol. 21, pp. 1061-1067 (DOI:10.1007/ s12540-015-5288-7)
49. A. Loida, and G. Schubert, (1991), Solidification of TRU Wastes in a Ceramic Matrix (EUR--13509). Commission of the European Communities (CEC), Vol. 23, 8, pp. xxx-xxx.
50. P. Presotto and V. Mymrine, (2012), Desenvolvimento De Novos Materiais Cerâmicos A Partir De Resíduo De Serpentinito E Argila Vermelha, 56° Congresso Brasileiro de Cerâmica 1° Congresso Latino-Americano de Cerâmica IX Brazilian Symposium on Glass and Related Materials 03 a 06 de Junho de 2012, Curitiba, PR, Brasil, pp. 923- 934.
51. S. Biff, and M. R. D. Silva, (2016), Use of Solid Waste from Sand Beneficiation Process in the Ceramic Tile Industry and its Influence on the Physical

Properties of the Ceramic Products, Cevista Matéria, Vol..21, pp.853-865 (10.1590/S1517-707620160004.0103).
52. V. M. Oversby, R. A. Van Konynenburg, W. E. GI.,Assley, and P. G. Curtis, (1993), (IMMt)Bilization In Ceramic Waste Forms of Tile L Residues from Treatment of Mixed Wastes / Scientific Basis for Nuclear Waste Management XVII Boston, MA November 29 - December 3, pp. 1-14.
53. A., Harker and J. Flintoff, (1983). Polyphase Ceramic and Glass-Ceramic forms for Immobilizing ICPP High-Level Nuclear Waste. MRS Online Proceedings Library (OPL), 26, 513. doi:10.1557/PROC-26-513
54. M. Marangoni, (2016), Dense and Porous Glass and Glass Ceramics from Natural and Waste Raw Materials, Archivio istituzionale della ricerca-Università di Padova, Doctoral Thesis. 2016, Providers: Archivio istituzionale della ricerca - Università di Padova
55. N.E. Toropkov, V.I. Vereshchagin, T.S. Petrovskaya, and N. S. Antonki, (2016). Influence of Synthesis Conditions on the Crystallinity Of Hydroxyapatite Obtained By Chemical Deposition, IOP Conf. Series: Materials Science and Engineering 156 012038 doi:10.1088/1757-899X/156/1/012038 (Materials and Technologies of New Generations in Modern Materials Science).
56. B. Jirajariyavej, P. Wanapirom, and C. Anunmana, (2018), Influence of Implant Abutment Material and Ceramic Thickness on Optical Properties, Journal Prosthet Dent, Vol. 119, pp. 819-825 (doi: 10.1016/j.prosdent.2017.05.015).
57. A. Şoim, M. Strîmbu, A.V Burde, B. Culic, D. Dudea, and C. Gasparik, (2018), Translucency and Masking Properties of Two Ceramic Materials for Heat-Press Technology, J Esthet Restor Dent, Vol. 30, pp. :E18-E23 (doi: 10.1111/jerd.12358).
58. Diana, S., Fauzan, R., Arahman, N., Razi, F., & Bilad, M. R. (2020). Synthesis and Characterization of Ceramic Membrane from Fly Ash and Clay Prepared by Sintering Method at Low Temperature. Rasayan Journal of Chemistry, Vol. 13, pp. 1335–1341 (https://doi.org/10.31788/RJC.2020.1335707).
59. S.-H. Ma, M.-D. Xu, X.-H. Wang, and X. Zhou, (2017), Challenges and Developments in the Utilization of Fly Ash in China, International Journal of Environmental Science and Development, Vol. 8, pp. 781-785. (doi: 10.18178/ijesd.2017.8.11.1057).
60. S. Maitra, (1999), Ceramic Products from Fly Ash Global perspectives, Fly Asia Utilization for Value Added Products Eds. B. Chatterjee, K. K. Singh & N. G. Goswami NML, Jamshedpur, pp. 32-3
61. E. Ercenka, U. Sena, G. Bayrakb, and S. Yilmaz, (2014), Glass and Glass-Ceramics Produced from Fly Ash and Boron Waste, Proceedings of the 3rd International Congress APMAS2013, April 2428, 2013, Antalya, Turkey, Vol. 125 (2014) Acta Physica Polonica A No. 2.
62. L. Carabba, S. Manzi, E. Rambaldi, G. Ridolfi, and M.C. Bignozzi, (2017), High-Temperature Behaviour of Alkali-Activated Composites based on Fly Ash and Recycled Refractory Particles, Journal of Ceramic Science and Technology, Vol. 08, pp. 377-388 (DOI: 10.4416/JCST2017-00047).

63. Y. Yang, F. Liu, Q. Chang, Z. Hu, Q. Wang, and Y. Wang, (2019), Preparation of Fly Ash-Based Porous Ceramic with Alumina as the Pore-Forming Agent, Ceramics, Vol. 2, pp. 286-295 (DOI: 10.3390/ceramics2020023).
64. P.W. Cleary, M.D. Sinnott, (2008), Assessing Mixing Characteristics of Particle-Mixing and Granulation Devices, Particuology, Vol. 6, pp. 419–444.
65. A. Sarkar, C.R. Wassgren, (2009), Simulation of a Continuous Granular Mixer: Effect of Operating Conditions on Flow and Mixing, Chemical Engineering and Science, Vol. 64 2672–2682.
66. A. Santomaso, M. Olivi, and P. Canu, (2004), Mechanisms of Mixing of Granular Materials in Drum Mixers under Rolling Regime, Chemical Engineering and Science, Vol. 59 pp. 3269–3280.
67. G. Léonard, F. Bertrand, J. Chaouki, P.M. Gosselin, (2008), An Experimental Investigation of Effusively as an Indicator of Powder Blend Uniformity, Powder Technology, Vol. 181, pp. 149–159.
68. H. Berthiaux, V. Mosorov, L. Tomczak, C. Gatumel, and J.F. Demeyre, (2006), Principal Component Analysis for Characterising Homogeneity in Powder Mixing using Image Processing Techniques, Chemical Engineering and Science, Vol. 45, pp. 397–403.
69. B. Chaudhuri, A. Mehrotra, F.J. Muzzio, and M.S. Tomassone, (2006), Cohesive Effects in Powder Mixing in a Tumbling Blender, Powder Technology, Vol. 165, pp. 105–114.
70. L.J. Bellamy, A. Nordon, D. Littlejohn, (2008), Effects of Particle Size and Cohesive Properties on Mixing Studied by Non-Contact NIR, Int. J. Pharm. Vol. 361, pp. 87–91.
71. H. Berthiaux, K. Marikh, C. Gatumel, (2008), Continuous Mixing of Powder Mixtures with Pharmaceutical Process Constraints, Chemical and Engineering Science, Vol. 47, pp. 2315–2322.
72. EMEA, The European Agency of the Evaluation of Medicinal Products. Note for guidance on process validation, 2001.
73. FDA, Food and Drug Administration. Process Validation. General Principles and Practices, U.S. Department of Health and Human Services, 2008.
74. D.W. Smith, (1983), Powder Metallurgy - Applications, Advantages and Limitations. American Society for Metals, Metal Park, Ohio.
75. D.T. Gethin, A. X., ArdIn, D. Y. Tran, and R.W. Lewis, (1994), Compaction and Ejection of Green Powder Compacts, International Journal of Powder metallurgy, Vol. 1, pp. xxx-xxx.
76. M. Gagne, (1997), Behavior of Powder, Mix Constituents during Cold and Warm Compaction. Advance in Powder Metallurgy could Particulate materials, 1
77. S. St-Laurent, F. Chagnon, and Y. Thomas, (2000), Study of Compaction and Ejection Properties of Powder Mixes Processed by Warm Compaction. Presented at the 2000 PM2TEC International Conference May 30-June 3, New York, USA.

78. A. Squire, (1947), Density Relationship of Iron-Powder Compacts, Transection AIME, Vol. 171, pp. 485-503.
79. W. Y. Park, C.S. Youn Y.W. Oh and C.J. Choi (2004), Characteristics of Fe Nanopowder Synthesized by Plasma Arc Discharge Process, Korean Journal of Materials Research, Vol. 14,
80. G. Thomson, (2003), Introduction to Nanotechnology, John Wiley & Sons, Inc. NY. USA
81. M. G. Fillabi A.Simchi and A. H., Kokabi (2008), Effect of Iron Particle Size on the Diffusion to Wrought Carbon Steels, Materials and Design, Vol. 29, pp. 411-417.
82. M. Delavari, A. Salarvand, A. Rahil, and F. Shahri, (2011), The Effect of Powder Metallurgy Process Parameters on Mechanical Properties of Micro and Nano-Iron Powder, International Journal of Engineering, Science and Technology, Vol. 3, pp. 86-94.
83. M. Alexandre, and P. Dubois, (2000), Polymer-layered Silicate Nanocomposites: Preparation, Properties and Uses of A New Class of Materials, Materials Science and Engineering R, Vol. 28, pp. 1-63.
84. H. P. Klugg, and L. E. Alexander, (1974), X-ray Diffraction Procedures; John Wiley & Sons: U.S.A.
85. D. Moore, and R. Jr. Reynolds, (1997), X-ray Diffraction and Identification and Analysis of Clay Minerals of Clay Minerals; Second Ed.; Oxford University Press, U. K.
86. FTIR Spectroscopy Attenuated Total Reflectance (ATR), www. perkinelmer. com.
87. Carl Zeiss Microscopy, (2004), Detection Principles based on GEMINI Technology.
88. Ghosh, M., Barman, A., De, S. K., and Chatterjee, S. (1998), Electrical Resistivity and Magnetoresistivity of Protonic Acid (H_2SO_4 and HCl)-Doped Polyaniline at Low Temperature, Journal of Applied Physics, Vol. 84, pp. 806–811.

3

High Resistance HCFC Slag Ceramic Materials

Muktikanta Panigrahi[1]*, Ratan Indu Ganguly[2] and Radha Raman Dash[3]

[1]*Department of Materials Science, Maharaja Sriram Chandra Bhanja Deo University, Keonjhar Campus, Odisha, India*
[2]*Department of Metallurgical Engineering, National Institute of Technology, Rourkela, Odisha, India*
[3]*CSIR-National Metallurgical Laboratory (NML), Jamshedpur, Jharkhand, India*

Abstract
This chapter discusses a novel highly resistant ceramic prepared from locally available high-carbon ferrochrome (HCFC) slag. After mixing ground fine HCFC slag particle of 240 mesh size with dextrin and 5% water, the paste is molded and sintered at four different sintering temperatures of 800 °C, 900 °C, 1000 °C, and 1200 °C. The sintered product is characterized with SEM, TEM, and XRD; and electrical properties evaluated by two-probe method. The effect of sintering temperature on electrical properties is discussed. The highest resistivity value of sintered HCFC slag is calculated to be 36.75×10^9 ohm cm. Microstructural analyses done by electron microscopy, i.e., SEM and TEM, reveal the presence of irregular shaped crystals dispersed in fused mass (i.e., glassy phase). The TEM micrograph shows a distinctly hexagonal shaped body with fused mass. The EDS analyses indicate the presence of elements, i.e., O, Si, Al, Fe, Cr, Mg, Ca, etc., both in sintered and raw HCFC slag; and. FTIR data signifies the presence of chemical groups within phases. Prominent phases like forsterite, spinel (magnesium aluminate), and clinopyroxene are identified by XRD analyses.

Keywords: High-carbon ferrochrome slag, sintered product, morphology, structure, resistivity

**Corresponding author*: muktikanta2@gmail.com

Muktikanta Panigrahi, Ratan Indu Ganguly and Radha Raman Dash (eds.) High Electrical Resistant Materials: Ferrochrome Slag Resource Ceramics, (83–96) © 2024 Scrivener Publishing LLC

3.1 Introduction

In the year 2014, 1.6 gigatons of steel were manufactured worldwide, which created 250 megatons of slag as waste [1]. Currently, tonnage of slags is expected to exceed 250 megatons.

Disposal of slag has become a major issue for steel industries and therefore poses a threat to our environment. This has led to active engineering research all over the world [2].

At present, most of the slag is used for making roads. However, it is also utilized for other purposes [3].

High-carbon ferrochrome (HCFC) is a common alloying material for the production of different grades of stainless steel. Global ferrochrome production in 2021 was *13.91 million tons* [4]. Ferrochrome slag is the residue generated during HCFC production, which is around 1.1–1.6 tons of FeCr produced, depending on feed materials [4].

Numerous studies are available in the literature about FeCr slag utilization for road and civil construction purposes such as coarse aggregate in concrete applications, pavement layers [5–9], partial substitution of fine aggregate sand in concrete [10, 11] and raw material in bricks [12]. Ferrochrome slag is used as aggregate in refractory castables for replacement of partly commercial aggregates. It is reported that 50 wt% of slag is used together with other ingredients, i.e., calcined bauxite [13, 14]. Such refractories show high compressive strength values ~60 MPa. This indicates the potentiality of ferrochrome slag in manufacturing of refractory products. Very limited results are available on the feasibility of using FeCr slag for high-temperature applications. The present study addresses this problem with a comprehensive investigation on castables refractory materials having the desired thermo-physico-mechanical properties. To the best of the authors' knowledge, there is a lack of information on ferrochrome slag utilization behavior of refractory materials at elevated temperatures. Mechanical and thermal properties of ferrochrome slag-based refractories are evaluated at high temperatures. This has yielded valuable information regarding high-temperature behavior of refractories produced from FeCr slag.

Refractories are ceramic materials which withstand a variety of harsh conditions. These include high temperatures, corrosive liquids and gases, abrasive wear, mechanical and thermal stresses [15]. They are indispensable for all high-temperature processes such as production of metals, cement, and glass ceramics. An enormous variety of refractories exist from which a particular category of refractories is selected depending on its availability and requirements in applications [16]. Refractory industries are dependent on the supply and availability of raw material and there has been a significant increase in material prices [17]. For the refractory industry, the need to secure raw materials at a competitive price has created a strong incentive for the exploration of alternative sources of raw materials to substitute virgin raw materials [18]. It is also equally important to develop a process that involves recycling of spent refractories [19].

In the current work, sintered high-carbon ferrochrome (HCFC) slag material is prepared via solid state. Materials are sintered at different sintering temperatures (i.e., 800 °C, 900 °C, 1000 °C, and 1200 °C, respectively). Microstructures of sintered product are examined with optical micrographs, TEM, and SEM attached with EDS. Phases of sintered HCFC slag materials are identified by X-ray diffraction technique; FTIR is done to identify the presence of different chemical groups in sintered HCFC slag samples; and electrical resistivity measurement is done by two-probe techniques.

3.2 Experiment Details

3.2.1 Materials and Chemicals

HCFC slag is collected from Shyam Ferro Alloys, India. Composition of raw HCFC slag is indicated in Table 3.1. Dextrin is used as binder (Merck, India). During green material preparation, a few drops of water (i.e., 6%) is used. Optical images of HCFC slag and dextrin are shown in Figure 3.1.

Table 3.1 Chemical compositions of HCFC slag (from Shyam Ferro Alloys, India).

	Used industrial waste
	HCFC slag
Constituents with percentages	FeO = 2.5–3.0 Al_2O_3 = 23.5–24.5 SiO_2 = 28–31 Cr_2O_3 = 4.5–5.4 MgO = 26.0–26.5 CaO = 7.5–8.5
Sources	Shyam Ferro Alloys, India
References	[20]

Figure 3.1 Optical image of HCFC slag (a) and Dextrin (b).

3.2.2 Sintered HCFC Slag Preparation

Preparation of sintered HCFC slag is achievable through the solid-state route [21, 22]. The following steps are adopted to obtain sintered HCFC slag materials.

Step 1: Processing of HCFC slag
As-collected HCFC slag is ground in a ball mill (for 5 h). It is sieved through a 240 mesh sieve. Sieved HCFC slag is dried in a heating oven at 120 °C (2 h). The purpose of heating sieved HCFC slag is to remove moisture. Such HCFC slag is called processed HCFC slag.

Step 2: Preparation of green pellet
Appropriate amount (10 g) of as-processed HCFC slag (i.e., 100% HCFC slag) is kept in a mortar with pestle; then 0.05% of dextrin is added to it as a binder. Both ingredients (HCFC slag and dextrin) are mixed in a mortar with pestle (1 h); and 6–10 mL of water is used to make the mixture pasty. Such pasty samples are known as green samples. Green sample is ready for preparation of pellets.

Pellets are made using a universal testing machine (UTM) and pelletizer. The assembled pelletizer is properly cleaned, washed, and dried before use. An appropriate amount of green sample is kept in the pelletizer. During compaction, 10 MPa pressure is applied to the sample contained in the pelletizer. It is left for five min for proper compaction. Then, the pelletizer is removed from the molding system and pellets are collected from the pelletizer. Such pellet is ready for sintering process.

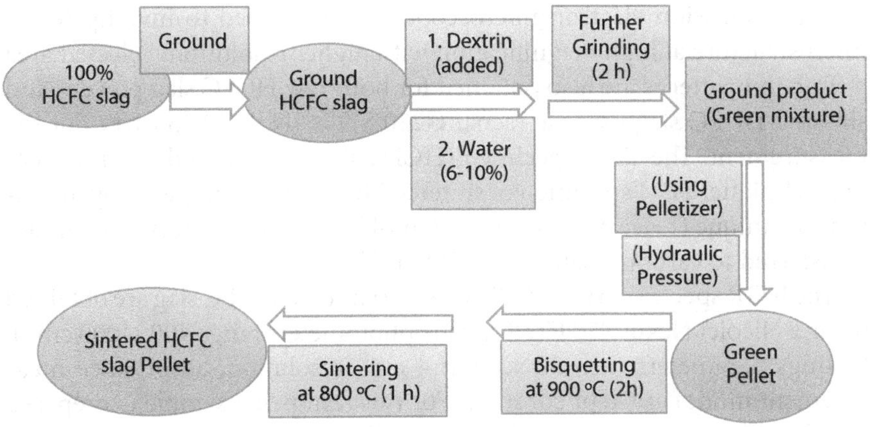

Figure 3.2 Flow chart showing the preparation of sintered HCFC slag-based materials.

Step 3: Preparation of sintered HCFC slag materials
Green pellet samples are put in a muffle furnace and bisquetted at 900 °C for 2 h to remove water and binder. Bisquetted pellets are sintered at 800 °C for 1 h. Dimensions (i.e., thickness and diameter) of sintered pellet are found to be 1.09 mm and 10 mm, respectively. The entire process is shown schematically in Figure 3.2. Sintered pellet(s) is ready for different characterizations for analyses.

A similar process is followed in other sets of sintered materials at different temperatures (i.e., 900 °C, 1000 °C and 1200 °C).

3.2.3 Test Methods

The XRD pattern is analyzed to identify different phases. A Philips PW-1710 Advance wide-angle X-ray diffractometer attached with a Philips PW-1729 X-ray generator is used. A Cu metal target is used to produce the X-ray; CuKα radiation is used for XRD analysis; and the XRD machine functions at 40 kV with tube current of 20 mA. Prepared pellet sample is put on a sample holder (made from quartz). Measurement is performed at room temperature. Sample is scanned trough diffraction angle (2θ) varying 10° to 70° with scanning speed of 2°/min.

Microstructural analyses are made with a scanning electron microscope (SEM, Carl Zeiss Supra 40). Sintered samples are gold-coated by sputtering technique. During the test, as an instrument parameter, the operating voltage is kept at 30 kV. EDS is performed along with microstructural analyses.

A transmission electron microscope (TEM) is used to investigate the microstructure at high magnification with higher resolution. Selected area diffraction patterns are also obtained for both raw HCFC slag powder and sintered HCFC slag using a Tecnai G220 S-TWIN TEM to carry out the measurement. The TEM specimen (HCFC slag) is prepared by drop casting technique. TEM specimen of sintered HCFC slag is prepared by microtone technique (Leica Mikrosysteme GmbH). Subsequently, specimens are transferred to carbon-coated Cu TEM grids.

The FTIR spectra of raw HCFC slag and sintered HCFC slag are obtained using a Nicolet Nexus 870 FTIR spectrophotometer (range 400–4000 cm^{-1}). Settings parameters (i.e., 50 scan at 4 cm^{-1} resolution, absorbance measurement mode) are kept constant. For this test, pellet sample is prepared with bulk amount of KBr powder and small amount of test sample through compression molding system with pelletizer. A background spectrum is collected before the test. Then, the pellet is put in an FTIR sample holder and data are collected.

For electrical resistivity measurement, sintered pellet samples are used. Resistivity value is estimated using the following equation [22]:

$$\rho = \left(\frac{R \times A}{l} \right) \tag{3.1}$$

Here, A is area of electrode $= \frac{\pi}{4} D^2$ (cm^2), R is resistance of prepared samples (Ω), l is thickness of sample, ρ is the resistivity of samples in Ω cm. For this measurement, a Keithyl source/multimeter is used.

3.2.4 Results and Discussion

The SEM images of processed HCFC slag (A) and HCFC slag sintered at 1200 °C (B) are shown in Figure 3.3. It is seen from the SEM image of HCFC slag (Figure 3.3a) that HCFC slag has irregular shaped crystals having some glassy texture. Irregular shaped crystals are dispersed in the background of fused glassy phase. Figure 3.3b shows the microstructure of HCFC slag sintered at 1200 °C. It is a fused mass structure. There is a distinct difference observed between the two microstructures (unsintered and sintered). This may happen due to the reaction between phases present in raw HCFC slag at 1200 °C.

The EDS analyses are done along with SEM studies (Figure 3.4a, b). The EDS profile of both the samples show the presence of elements such as O,

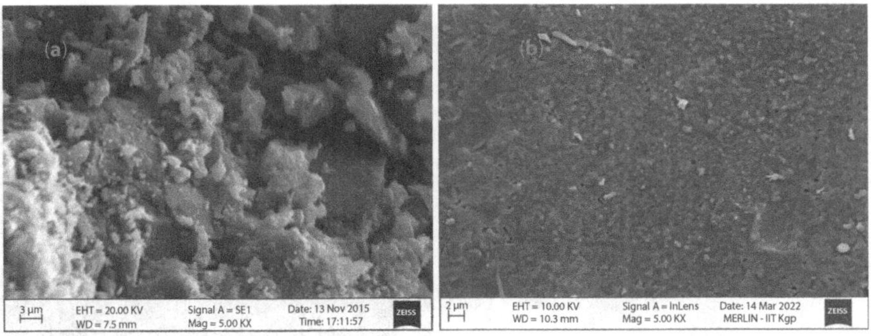

Figure 3.3 SEM image of processed HCFC slag (a) and HCFC slag sintered at 1200 °C (b).

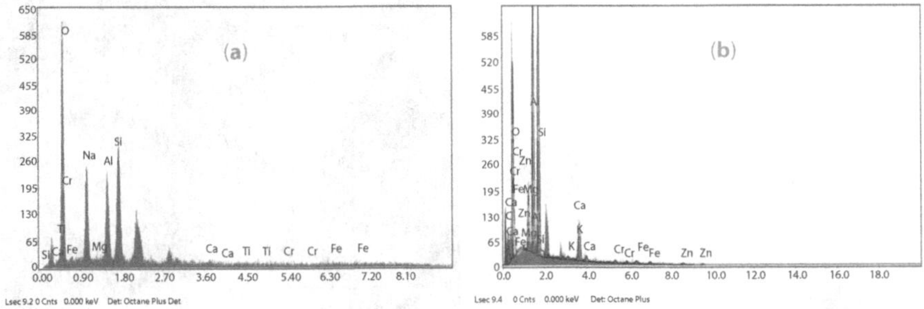

Figure 3.4 EDS spectrum of processed HCFC slag (a) and HCFC slag sintered at 1200 °C (b).

Al, Si, Ti, Fe, Cr, Ca, Zn, and Mg. Also, Ca, Zn, and Mg elements are present in trace amounts. The EDS results are corroborated with XRD results.

Figure 3.5a shows the TEM image of HCFC slag. Some fused crystals of HCFC slag are dispersed in the matrix, which are crystalline in nature.

The diffraction pattern of raw HCFC slag is shown in Figure 3.5b. Here, bright spots are concentrated with lesser spots present around the circumference of the circle. These spots support the presence of crystalline phase of raw samples.

Figure 3.5c is a TEM photograph of sintered HCFC slag. Some fused crystals of sintered HCFC slag are dispersed in the matrix, which are also crystalline in nature.

The diffraction pattern of sintered HCFC slag is shown in Figure 3.5d. Here, bright spots are concentrated with lesser spots spread around the circumference of the circle. Therefore, it also indicates the presence of crystalline phase in sintered HCFC slag samples.

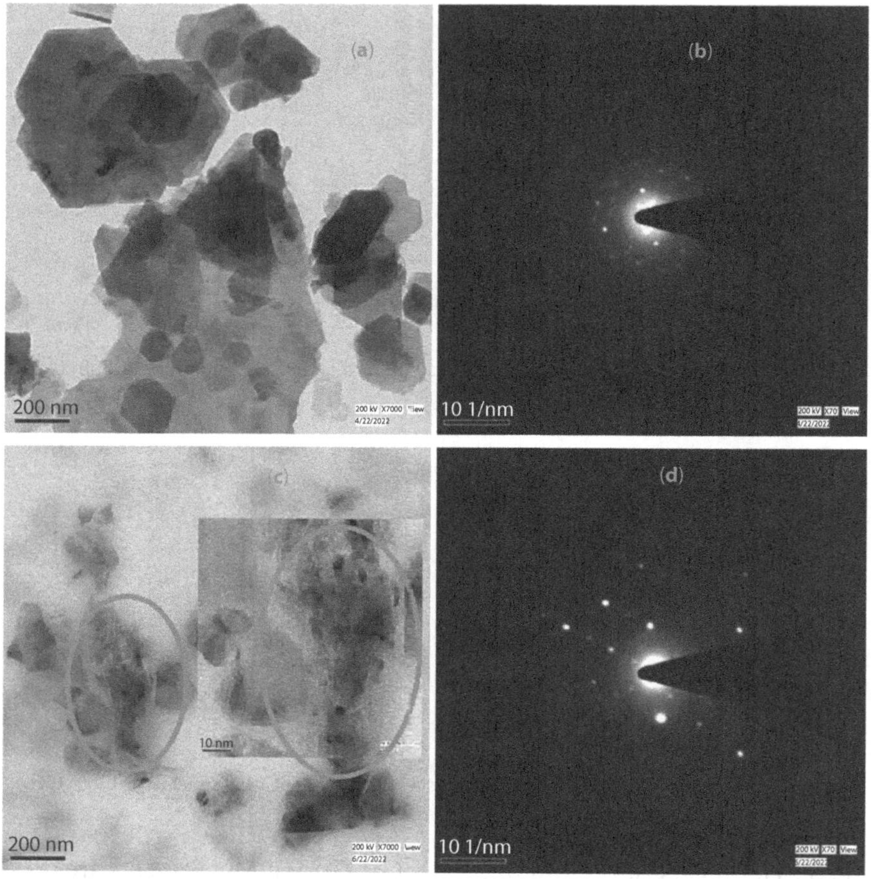

Figure 3.5 TEM images and selected area electron diffraction (SAED) pattern of HCFC slag (a & b) and sintered HCFC slag at 1200 °C (c & d).

Figure 3.6 displays the XRD pattern of raw HCFC slag and HCFC slag sintered at 1200 °C. In the XRD pattern (see Figure 3.6a), various prominent phases (i.e., forsterite ($MgSiO_4$), wollastonite ($CaSiO_3$), magnesium aluminate (spinel~$MgAl_2O_4$), and chromium-bearing phases [($Na_{45}Ca_{55}$)($Cr_{45}Mg_{55}$)$52O_2$] are identified. A few other phases, such as hematite (Fe_2O_3), silica (SiO_2), alumina (Al_2O_3), and mulite phase ($3Al_2O_3.2SiO_2$), are identified at different Bragg's angles [22].

Figure 3.6b shows the XRD pattern of sintered HCFC slag, which is sintered at 1200 °C. Various phases are observed in the spectrum (see Figure 3.6b) with different intensities. These phases are forsterite ($MgSiO_4$), wollastonite ($CaSiO_3$), magnesium aluminate phase (spinel~$MgAl_2O_4$), chromium-bearing phases [($Na_{45}Ca_{55}$)($Cr_{45}Mg_{55}$)$52O_2$], hematite (Fe_2O_3), silica (SiO_2), alumina (Al_2O_3), and mullite phase ($3Al_2O_3.2SiO_2$). The intense

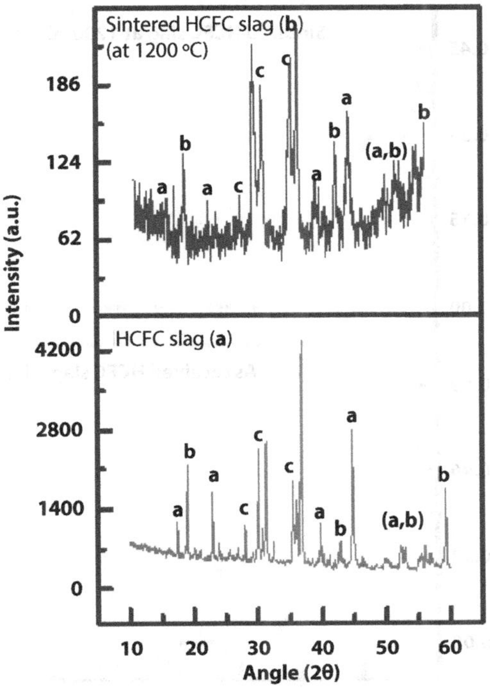

Figure 3.6 XRD pattern of raw HCFC slag (a) and 1200 °C sintered HCFC slag (b). [Note: a = Forsterite, b = Spinel (magnesium aluminate), c = Clinopyroxene]

peak is found at 36° (2Θ) in the XRD pattern (Figure 3.6) and the phase is oxide of chromium-bearing phases [$(Na_{45}Ca_{55})(Cr_{45}Mg_{55})52O_2$] (ICSD analysis). Magnesium aluminate phase is observed at 19°, 35° and 60°, respectively. There is a marked difference in intensity value before and after sintering (Figure 3.6a, b). This signifies that some reaction happened during sintering.

Figure 3.7a shows FTIR absorption bands of raw HCFC slag. Stretching band at 620 cm^{-1} matches with the Si-O-Al band. Stretching band is found at 1079 and 1648 cm^{-1}. This band may be accredited to Si-O-Si asymmetric band and H-O-H bending band, respectively. Other stretching bands are 3438, 2237, 983, and 522. Corresponding bands are OH group, carbonate group, Si-O group, and O-Si-O group, respectively [23–31].

The FTIR spectrum of HCFC slag sintered at 1200 °C is shown in Figure 3.7b. As can be seen in this figure, broad stretching bands are found around 3426 cm^{-1}. The band arises due to the stretching vibrations of O-H bonds and H bending vibrations H-O-H of interlayer adsorbed

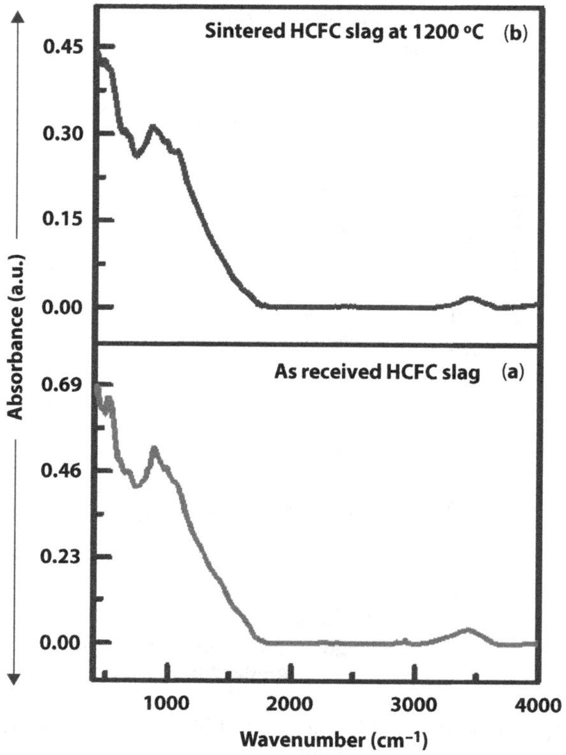

Figure 3.7 FTIR spectra of processed HCFC slag (a) and HCFC slag sintered at 1200 °C (b).

H_2O molecule. A C=O stretching band is identified at 2270 cm^{-1} and indicates the occurrence of carbonate groups. This is mainly attributed to the presence of chemisorbed CO_2 in sintered HCFC slag. Si-O and O-Si-O bands are also found at 1071 cm^{-1} and 495 cm^{-1}. They indicate the presence of silicate groups. The presence of Al^{3+}-O^{2-} bonds are also observed near 858 cm^{-1}. Fe-O stretching band is also observed at 409 cm^{-1} [23–31].

Table 3.2 indicates the resistivity values of as-processed HCFC slag, HCFC slag sintered at 800 °C, 900 °C, 1000 °C and 1200 °C, respectively. From the estimated resistivity data (see Table 3.2), the highest resistivity value is found to be 36.75×10^9 Ohm cm (for HCFC slag sintered at 1200 °C) and the lowest resistivity value is 3.25×10^9 Ohm cm (for HCFC slag sintered at 900 °C). It is concluded that the highest resistivity value is due to formation of mullite phases, magnesium aluminate phase, and oxides of chromium-based complex phase [22].

Table 3.2 Resistivity value of sintered HCFC slag samples.

S. no.	Sample ID	Sintered temperature	Resistance value (GΩ)	Resistivity value (Ohm.cm)
1	HCFC slag	---	---	---
2	Sintered HCFC slag	800 °C	Not estimated	Not estimated
3	Sintered HCFC slag	900 °C	0.23	3.25×10^9
4	Sintered HCFC slag	1000 °C	0.43	6.08×10^9
5	Sintered HCFC slag	1200 °C	2.6	36.75×10^9

3.2.5 Conclusions

- Sintered HCFC slag materials are successfully prepared via the sintering process. With increase in sintering temperature from 800 °C to 1200 °C, it is seen that electrical resistance value increases.
- Phases in sintered material and raw HCFC slag are identified with an X-ray diffractometer. Identified phases are forsterite, spinel (magnesium aluminate), and clinopyroxene.
- Microstructure of raw HCFC slag shows irregular shaped crystals, whereas sintered product indicates fused glassy phase having smooth texture.
- Different elements present in sintered material are identified by EDS analyses. Also, TEM analysis revealed the presence of hexagonal shaped crystals in raw HCFC slag. TEM analysis of the microstructure of sintered slag shows fused crystals in the matrix.
- FTIR spectra indicate significant changes. Changes in the peaks are due to formation of new bonds within the phases during sintering of raw HCFC slag.
- The maximum electrical resistance (resistivity) of sintered slag (at 1200 °C) is found to be 2.6 gigaohm (36.75×10^9 Ohm cm).

Acknowledgments

The authors would like to thank Prof. Munesh Chandra Adhikary, PG Council Chairman, Fokir Mohan University, for his invaluable guidance, advice, and constant inspiration throughout the entire program. They also would like to thank Mr. Mukteswar Mohapatra of Fokir Mohan University for his support. The authors convey their sincere thanks to the management of GIET University, Gunupur, Rayagada, Odisha, India for providing lab facilities to do the research work. They also wish to thank staff of CRF IIT Kharagpur for providing testing facilities.

References

1. N.S. Pandian, C. Rajasekhar, and A. Sridharan, (1998), Studies of the Specific Gravity of Some Indian Coal Ashes. J. Test. Eval., Vol. 26, pp. 177-xxx.
2. N.S. Pandian, (2004), Fly Ash Characterization with Reference to Geotechnical Applications. J. Indian Inst. Sci., Vol. 84, pp. 189-xxx.
3. M. Ahmaruzzaman, (2010), A Review on the Utilization of Fly Ash, Prog. Energy Combust. Sci., Vol. 36, pp. 327-xxx.
4. L. Wang, G. Hu, F. Lyu, T. Yue, H. Tang, H. Han, Y. Yang, R. Liu,and W. Sun, (2019), Application of Red Mud in Wastewater treatment. Minerals, Vol. 9, pp. 281-xxx.
5. S. Sushil, V.S. Batra, (2008), Catalytic Applications of Red Mud, an Aluminium Industry Waste: A Review. Appl. Catal. B Environ., Vol. 81, pp. 64.
6. C.L. Goldsmith, A. Malczcwski, J.J. Yao, S. Chen, J. Ehmk, D.H. Hinzel, (1999), RFMEMS-Based Tunable Filters. Int. J.RF Microwave CAE, Vol. 9, pp. 362-xxx.
7. G. Subramanyam, F.V. Keuls and F. A. Miranda, (1998), Novel K-band Tunable Microstrip v-band Pass Filter using Thin Film HTS/Ferroelectric/ Dielectric Multilayer Configuration. IEEE Microwave Guided Wave Let, Vol. 8, pp. 78-xxx.
8. A. Tombak, J.P. Maria, F.T. Ayguavives, Z. Jin, G.T. Stauf, A.I. Kingon, and A. Mortazawi, (2003), Voltage-Controlled RF Filters Employing Thin Film Barium-Strontium Titanate Tunable Capacitors. IEEE Trans. Microwave Theory and Tech., Vol. 51, pp. 462-xxx.
9. D. Goski and M. Lambert, (2019), Engineering Resilience with Precast Monolithic Refractory Articles, Int J Ceramic Engg. Sci., Vol. 1, pp. 169-177.
10. L. Li and Y. Li, (2017), Development and trend of ceramic cutting tools from the perspective of mechanical processing, IOP Conf. Series: Earth and Environmental Science, Vol. 94, pp.1-6.
11. M. Zhu, R. Ji, Z. Li, H. Wang, L.L. Liu, and Z. Zhang, (2016), Preparation of Glass Ceramic Foams for Thermal Insulation Applications from Coal

Fly Ash and Waste Glass, Construction and Building Materials, Vol. 112, pp. 398-405.
12. H. E. Exner, and E.Arzt, (1996), Sintering Processes. In Physical Metallurgy, Vol.3, ed. R. W. Cahn and P. Haasen. 4th ed. Elsevier Science, Amsterdam, pp. 2628-2662.
13. W.V. Siemens, (1966), Inventor and Entrepreneur: Recollections of Werner Von Siemens". London, England.
14. T. Tanaka and T. Imai, (2013), Advances in Nanodielectric Materials over the past 50 years, IEEE Electrical Insulation Magazine, Vol. 1, pp. 10-23.
15. T. Tanaka, (2005), Dielectric Nanocomposites with Insulating Properties, IEEE Transactions on Dielectrics and Electrical Insulation, Vol. 12, pp. 914-928.
16. E.C. Nzenwa, and A.D. Adebayo, (2019), Analysis of Insulators for Distribution and Transmission Networks, American Journal of Engineering Research (AJER), Vol.8, pp-138-145.
17. R. Kulke, G. Möllenbeck, C. Günner, P. Uhlig, K.H. Drue, S. Humbla, J. Müller, R. Stephan, D. Stöpel, J.F. Trabert, G. Vogt, M.A. Hein, A. Molke, T. Baras, A.F. Jacob, D. Schwanke, J. Pohlner, A. Schwarz, and G. Reppe, (2009), Ceramic Microwave Circuits for Satellite Communication, Journal of Microelectronics and Electronic Packaging Vol. 6, pp. 27-31
18. J. Nath, D. Ghosh, J.P. Maria, A.I. Kingon, W. Fathelbab, D. Paul, F.Z. Michael, and B. Steer, (2005), An Electronically Tunable Microstrip Band Pass Filter using Thin-Film Barium-Strontium Titanate (BST). IEEE Trans. Microwave Theory and Tech., Vol. 53, pp. 2707-xxx.
19. L.Y.V. Chen, R. Forse, D. Chase, and R.A. York, (2004), Analog Tunable Matching Network using Integrated Thin-Film BST Capacitors. IEEE MTT-S Digest, Vol. 1, pp. 261-xxx.
20. F.H. Weel, F. Malek, S. Sreekantan, A.U. Al-Amani, F. Ghanil, and K.Y. You, (2011), Investigation of the Characteristics of Barium Strontium Titanate (BST) Dielectric Resonator Ceramic Loaded on Array Antennas. Prog. Electromag. Research, Vol. 121, pp. 181-xxx.
21. M.K. Panigrahi, P. Kumar, B. Barik, D. Behera, S.K. Mohapatra, and H. Jha, (2016), Frequency Dependency of Developed Dielectric Material from Fly Ash: An Industrial Waste. In. 20th National Conference on Nonferrous Minerals and Metals, 8-9th July 2016; Eds. Rakesh Kumar, K.K.Sahu & Abhilash, Vol. xxx, pp. 143-xxx.
22. M.K. Panigrahi, (2021), Investigation of Structural, Morphological, Resistivity of Novel Electrical Insulator: Industrial Wastes, Bull. Sci. Res., Vol. 3, pp. 51-58.
23. B.J. Saikia, G.K Rao, and P. Sarathy, (2010), Fourier Transform Infrared Spectroscopic Characterization of Kaolinite from Assam and Meghalaya, North Eastern India, J. Mod. Phys., Vol. 1, pp. 206-xxx.
24. R.P. dos Santos, J. Martins, C. Gadelha, B. Cavada, A.V. Albertini, F. Arruda, M. Vasconcelos, E. Teixeira, F. Alves, J.L. Filho, and V. Freire, (2014), Coal Fly Ash Ceramics: Preparation, Characterization, and Use in the Hydrolysis of Sucrose. Scientific World Journal, Vol. 2014, pp. 01-xxx.

25. A. Jose, M. R. Nivitha, M. Krishnan, and R. G. Robinson, (2020), Characterization of cement stabilized pond ash using FTIR spectroscopy, Construction and Building Materials, Vol. 263, pp.120136-xxx (DOI: 10.1016/j.conbuildmat.2020.120136).
27. Y. Liu, F. Zeng, B. Sun, P. Jia, and I. T. Graham, (2019), Structural Characterizations of Aluminosilicates in Two Types of Fly Ash Samples from Shanxi Province, North China, Minerals, Vol. 9, pp. 358-16 (doi:10.3390/min9060358).
28. P. Rożek, M. Król, and W. Mozgawa, (2018), Spectroscopic Studies of Fly Ash-Based Geopolymers, Spectrochim Acta - Part A Mol Biomol Spectrosc, Vol. 198, pp. 283-289 (https://doi.org/10.1016/j.saa.2018.03.034).
29. S. Kumar, F. Kristály, and G. Mucsi, (2015), Geopolymerisation Behaviour of Size Fractioned Fly Ash, Adv Powder Technol, Vol. 26, pp. 24-30 (https://doi.org/10.1016/j.apt.2014.09.001).
30. Z. Kledyński, A. Machowska, B. Pacewska and I. Wilińska, (2017), Investigation of hydration products of fly ash-slag pastes, Journal of Thermal Analysis and Calorimetry, Vol. 130, pp. 351-363 (https://doi.org/10.1007/s10973-017-6233-4).
31. D. Bondar and R. Vinai, (2022), Chemical and Microstructural Properties of Fly Ash and Fly Ash/Slag Activated by Waste Glass-Derived Sodium Silicate, Crystals, Vol. 12, pp. 913-12 (https:// doi.org/10.3390/cryst12070913).

4

High Resistant HCFC Slag/Kaolin Ceramics

Muktikanta Panigrahi[1]*, Ratan Indu Ganguly[2] and Radha Raman Dash[3]

[1]*Department of Materials Science, Maharaja Sriram Chandra Bhanja Deo University, Keonjhar Campus, Odisha, India*

[2]*Department of Metallurgical Engineering, National Institute of Technology, Rourkela, Odisha, India*

[3]*CSIR-National Metallurgical Laboratory (NML), Jamshedpur, Jharkhand, India*

Abstract

Development of value-added product from waste material has become a research trend for R&D scientists. Utilizing such waste protects the environment and its commercial use also strengthens the global economy. One such waste obtained from the ferrochrome industry is high-carbon ferrochrome (HCFC) slag.

Composite is made by mixing HCFC slag with kaolin. The main constituents of HCFC slag are SiO_2, Al_2O_3, Cr_2O_3, FeO, FeO, CaO and MgO. Before sintering, the composite mixture is compacted at a pressure of 10 MPa. Then, it is sintered at four different temperatures of 800 °C, 1000 °C, 1100 °C, and 1200 °C for 1 h to develop a ceramic composite. After determining the electrical resistance value (resistivity value) and hardness value of the products, their microstructures are characterized. The occurrence of chemical groups of sintered products are also investigated.

Ceramic developed with 70% HCFC slag and 30% kaolin has enabled acceptable resistance properties. Maximum electrical resistivity value of the composite (70% HCFC slag and 30% kaolin) sintered at 1200 °C is found to be 18.65×10^{11} Ω cm. The Vickers hardness value determined by the nanoindentation method is found to be 865.72. X-ray diffraction (XRD), Fourier-transform infrared (FTIR) spectroscopy, optical microscopy, scanning electron microscopy (SEM), and transmission electron microscopy (TEM) are used to characterize developed materials.

By controlling composition and sintering temperature, it is possible to form phases such as mullite phase, spinel phases, forsterite phase, chromium-based phases, and clinopyroxene phase, which enhance strength and electrical resistance of the composite. Such phases are explored by XRD analysis.

**Corresponding author*: muktikanta2@gmail.com

Muktikanta Panigrahi, Ratan Indu Ganguly and Radha Raman Dash (eds.) High Electrical Resistant Materials: Ferrochrome Slag Resource Ceramics, (97–116)

Keywords: Industrial wastes, HCFC slag, kaolin, composite, XRD, phases, morphology, hardness, resistivity

4.1 Introduction

The clean and economical disposal of slags containing heavy metals is essential for an eco-friendly society. Ferrochrome slag (FS) and tundish slag (TS) are two typical slags with high content of heavy metals. Ferrochrome slag is a bulky by-product of ferrochrome industries. Global FS production was approximately 11.7 million tons in 2017, which is expected to increase by 3.6% annually [1]. Recycling and efficient usage of large amounts of FS are posing challenges for ferrochrome producers because of its high content of Cr_2O_3.

Research on the utilization of FS has been broadly classified under the following categories;

1. Refractory applications, such as refractory castables [2].
2. Applications in the construction of roads and buildings [3] as aggregates in concrete [4].
3. Applications in treatment of hazardous hexavalent Cr [5], i.e., removal of hexavalent Cr from jigging water streams.

Tundish slag is discharged from tundish after continuous casting of steel. A few studies have reported on the reuse of TS. According to an investigation performed at Shandong Steel in China, approximately 2000 tons of TS are released from a workshop every year and accumulate in landfills without being reused. A high content of Mn (18.69 wt%) was detected in TS in 2015, and similar results are reported by others [6].

The ceramic industry is the third-largest silicate industry after cement and concrete industries. For this, slag is used in bulk quantities by manufacturers of ceramic products. Uses of these raw materials for ceramic production have improved quality of products. Ceramics can withstand high temperature and have good creep resistance. Therefore, they are used as lining materials on mold walls where molten liquid metal or alloys solidify (1200–1700 °C) to form ingots. Such lining cannot be met with cement or concrete. They are used to manufacture building ceramics (wall tiles and floor tiles) in large quantities. They have comparatively lower environmental impact. However, results of research on reuse of FS and TS as raw materials for ceramic industries have rarely been reported.

Recently, Zhao *et al.* [7–10] synthesized ceramics from solid waste containing high amount of CaO and Fe_2O_3. These wastes include steel slag, tailings and red mud. Ceramics made from these wastes primarily contain pyroxene group minerals instead of quartz or mullite. Pyroxene ceramics exhibit mechanical properties having high bending strength (107–143 MPa). This value is significantly higher than the strength properties of porcelain (ISO 13006: 2012: 35 MPa) made from other raw materials [7-10]. Ceramics made from red mud retain pyroxene and anorthite. This should reduce leaching rate in comparison to those ceramics prepared from material with low Na^+ ion [11]. However, only a few studies have been conducted on the Cr/Mn leaching behaviors of pyroxene ceramics. Cr and Mn are toxic elements which influence the neurological and muscular functions of human bodies. Therefore, it is necessary to stabilize chromium in stainless steel slag by converting to spinel structure (AB_2O_4), where A is divalent element and B is trivalent element (e.g., $MgCr_2O_4$) [12, 13]. Ionic radii of Mn^{2+}, Fe^{2+}, and Zn^{2+} have different types of spinel structure ($Mn_{0.66}Zn_{0.25}Fe_{2.09}O_4$). These possess high resistance in acid environment [14]. Therefore, pyroxene/spinel ceramics are sustainable materials for processing ceramics. They will have good mechanical properties with least leaching problem due to stabilization of Cr and Mn.

For preparation of insulators having good insulation properties, it is essential that composite is made from HCFC slag by mixing with clay material (i.e., kaolin).

Clays are abundantly available in nature. Clay has many properties which justifies its many technological applications (i.e., civil construction and oil wells, chemical, white ceramic, food, drugs and cosmetics, filler for polymers, etc.). Clays are used as minor components to prepare composites [15]. The variety of uses of clay is due to its chemical, mineralogical, technological and economic aspects [16]. Commercially, the process is viable. Usually, clay minerals contain kaolinite, bentonite, smectite, diatomite and Fuller's earth. Also, they substitute activated carbon because of their good sorption properties [17]. Structurally, they constitute layered aluminosilicate minerals. Crystallographic features of layers consist of TO_4 tetrahedra ($T = Si^{4+}$, Al^{3+}, etc.) and MO_6 octahedra ($M = Al^{3+}$, Fe^{3+}, etc.). Layers belong in 1:1 family of clay minerals and are held together by hydrogen bonds. They are formed between hydroxyl groups attached to coordinately unsaturated oxygen sites of the MO_6 side of each layer and oxygen atoms terminating the opposite SiO_4 side of the next layer [18]. Structural units of clay consist of silica sheet (formed of silica tetrahedra) and the octahedral units formed with octahedrally coordinated cations with oxygens or hydroxyls octahedra) [19]. Natural clay contains impurities (i.e., quartz, calcite, feldspar, mica and organic matter, hydrated iron oxide, ferrous carbonate and pyrite). Physical properties are affected by particle size, shape, cation exchange capacity and the type of impurities present. Usually, the size of clay particles varies from micrometer to nanometer scale.

Kaolinite is a metamorphic rock which is formed under different environmental influences, i.e., weathering, diagenetic, hydrothermal, etc. Kaolinite is an abundant mineral. Clay particles have different shapes with various aspect ratios. For all the above reasons, kaolinite plays a vital role in controlling chemical and physical characteristics of the products. Concerning their properties, the degree of crystallinity, concentration of impurities, and particles size distribution will vary if they are procured from different areas [20].

Ceramics are used for different purposes, viz., refractories [21], cutting tools [22], thermal and electrical insulations [23], etc. The preparation process of ceramics requires adoption of high pressure followed by sintering at an elevated temperature. Performances of ceramics are related to the presence of crystalline phase(s) in sintered products. Therefore, composition of raw materials, process parameters, and sintering temperature play key roles in achieving desired phases in the sintered products. Formation of mullite phase has been reported to be beneficial in achieving high resistance of the sintered materials. The mullite phase is formed at a firing temperature of 1150–1200 °C [24].

Ceramics were used as electrical insulators in 1850 for construction of electrical air lines [25]. Distinctive properties of ceramics (i.e., mechanical strength, high-power dielectric strength, and corrosion resistance) have made them indispensable in areas of electric line construction [26, 27]. Looking to the current demands, scientists are motivated to develop insulators having good properties with low cost [28].

In the present investigation, composite is made from HCFC slag and Kaolin material sintered at four different temperatures, i.e., 800 °C, 1000 °C, 1100 °C and 1200 °C. Raw materials mixed in requisite proportions and pellets are made from them by the usual process of compaction and bisquetting followed by sintering. Sinter products are characterized by different techniques, i.e., optical microscopy, SEM with EDS, TEM, FTIR, XRD, and electrical resistivity. Phases present in raw materials have also been identified by X-ray diffractometry.

4.2 Experiment Details

4.2.1 Materials and Chemicals

The HCFC slag was procured from Shyam Ferro Alloys, Durgapur, India. Kaolin, dextrin $(C_6H_{10}O_5)_n \cdot xH_2O$) was procured from Merck, India. During the composite preparation, a few drops of water (i.e., 6%) were used. Compositions of processed HCFC slag and kaolin are indicated in Table 4.1. Optical images of processed slag (HCFC slag), kaolin, and dextrin are shown in Figure 4.1.

Table 4.1 Constituents with percentage (%) of processed HCFC slag and kaolin.

	Compositions (%)												
S. ID	**Fe_2O_3**	**Al_2O_3**	**SiO_2**	**FeO**	**Cr_2O_3**	**Na_2O**	**MgO**	**K_2O**	**CaO**	**TiO_2**	**BaO**	**LOI**	**Ref**
HCFC slag	2.5-3.0	23.5-24.5	28-31	---	4.5-5.4	---	26-26.5	---	7.5-8.5	---	---	---	[29]
CC	---	37.66	44.80	1.54	---	---	T	---	0.50	0.60	---	14.33	[30]

Note: S. ID = Sample ID, HCFC slag = High-carbon ferrochrome slag, CC = Kaolin.

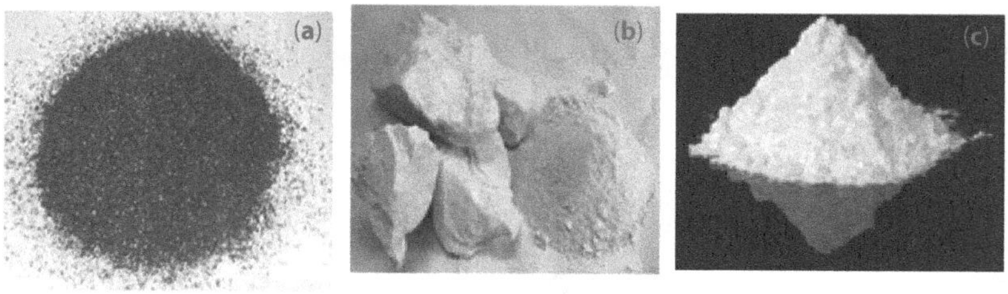

Figure 4.1 Optical images of HCFC slag (a), kaolin (b) and dextrin (c).

4.2.2 HCFC Slag/Kaolin Composite Preparation

Preparations of HCFC slag/kaolin materials are achievable through solid-state route [30, 31]. The steps adopted for material preparation are described below.

Step 1: Preparation of processed HCFC slag
Collected HCFC slag is ground in a ball mill (5 h). It is sieved through 240 meshes. Sieved HCFC slag is dried in a heating oven at 120 °C (2 h). The purpose of heating sieved HCFC slag is to remove the moisture from it. Such HCFC slag is called processed HCFC slag.

Step 2: Preparation of green pellet
An appropriate amount of processed HCFC slag is placed in a mortar with pestle. Requisite amount of kaolin is added to the mortar. They are ground together to make the mixture pasty. Small amount of dextrin (i.e., 0.05%) is mixed further with the mixture, and then grinding is continued for ½ h to ensure uniformity in the mixture. Dextrin has been used as binder. Small amount of water is further used to make mixture pasty. Pellets of mixture are then cast during their mixing. Such mixture samples are known as green samples. Samples are ready for preparing pellet.

Pellets of mixture are then cast in a pelletizer. Before pelletizing, components of pelletizer are cleaned, washed, and dried before use. An appropriate amount of powder mixture is kept in the pelletizer and components are reassembled. The assembly is then placed on a universal testing machine (UTM). A compression pressure of 10 MPa is applied to the pealletizer. Under compaction, pelletizer is kept for five minutes for their proper compaction. Then, pelletizer is removed from the UTM and is disassembled. Pellet is removed from molding system. A number of pellets are prepared with different composition of slag and kaolin in a similar manner (Table 4.2). An average dimension (thicknesses and diameter) of pellet has been

Table 4.2 Compositions (%) of HCFC slag/CC composite preparation.

S.no.	Individual composition name		
	Sample ID	**HCFC slag**	**Kaolin (CC)**
1	T1	**50**	**50**
2	T2	**60**	**40**
3	T3	70	30
4	T4	80	20
5	T5	90	10

noted. This is estimated to be 1.1 mm (thickness) and 10 mm (diameter). Pellet is ready for using sintering process.

Step 3: Sintering of pellets (HCFC slag/kaolin-based materials)
Green pellet samples are kept in a muffle furnace, which is heated at 900 °C (2 h) to remove water and binder. This is called the bisquetting process. Bisquetted pellet(s) are then sintered at a temperature of 1200 °C (1 h). The average dimensions (i.e., thickness and diameter) of sintered pellet are noted to be 1.09 mm thickness and 10 mm diameter. The entire process is shown schematically in Figure 4.2. Sintered pellet(s) has undergone different characterization processes.

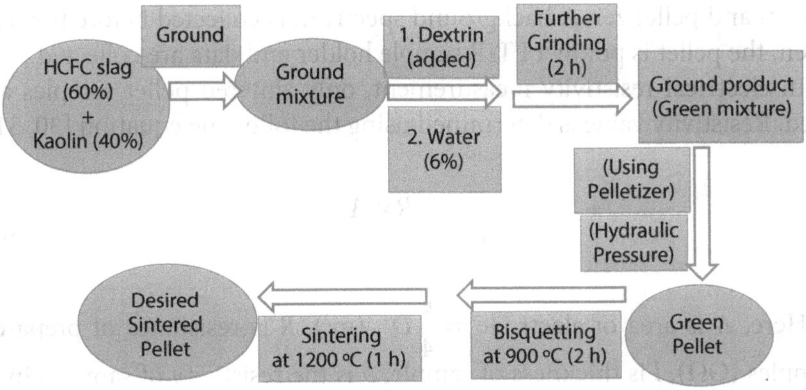

Figure 4.2 Flow chart showing the preparation of sintered HCFC slag/kaolin-based materials.

4.2.3 Test Methods

X-ray diffraction (XRD) analyses are used to identify different phases. For carrying out the analyses, a Philips PW-1710 Advance wide-angle X-ray diffractometer attached with a Philips PW-1729 X-ray generator has been used. For analyses, CuKα radiation is used. Sintered pellet sample is placed on a quartz sample holder. Sample is scanned through a Bragg's angle varying 20° to 70° with scanning speed 2°/min.

The optical image of sintered HCFC slag is examined. Vickers hardness of sample has been measured by a nanoindentation tester NHT (Anton Paar, Switzerland).

Morphological analyses are made by a Carl Zeiss Supra-40 scanning electron microscope (SEM). Sintered samples are gold-coated (sputtering technique) before being examined. Along with morphological studies, EDS analyses have been made.

Raw HCFC slag powder and sintered HCFC slag are examined by a transmission electron microscope (TEM, 5022/22, Tecnai G2 20 S-Twin, Czech Republic). TEM specimen of HCFC slag is prepared by drop casting technique, and TEM specimen of sintered HCFC slag is prepared by microtone technique (Leica Microsystems GmbH, A-1170). Then, specimens are transferred to carbon-coated Cu TEM grids.

The FTIR spectra of raw materials (HCFC slag and kaolin) and sintered HCFC slag/kaolin samples are obtained. For FTIR analyses, Thermo Nicolet Nexus 870 spectrophotometer (range 400–4000 cm^{-1}) is used. Settings parameters (i.e., 50 scan at 4 cm^{-1} resolution, absorbance measurement mode) are kept constant. For this test, all test samples are powdered and pellet samples are prepared using KBr through a compression molding system and pelletizer. A background spectrum is collected before the test. Then, the pellet is put in a FTIR sample holder and data are collected.

In electrical resistivity measurement, only sintered pellet samples are used. Resistivity value is determined using the following equation [30, 31]:

$$\rho = \left(\frac{R \times A}{l} \right) \tag{4.1}$$

Here, A is area of electrode $= \frac{\pi}{4} D^2$ (m^2), R is resistance of prepared samples (GΩ), l is thickness of sample, ρ is the resistivity of samples in Ω cm. For this measurement, a multimeter is used.

4.2.4 Results and Discussion

The aim of this work is to develop an insulator having the best electrical insulation property. Therefore, the effect of sintering temperatures on insulating property is measured. Both composition and sintering temperature are varied in order to get maximum resistivity value for a particular composition of composite and sintering temperature. Composition of composites designated as T1, T2, T3, T4, and T5 are shown in Table 4.2. Tables 4.3, 4.4, 4.5, and 4.6 show resistance value (resistivity value) of composites having five different compositions and sintered at a temperatures of 800 °C, 1000 °C, 1100 °C and 1200 °C, respectively.

Table 4.3 shows resistance value (resistivity value) for five compositions of composites at a sintering temperature of 800 °C. The highest resistance value (resistivity value) is found for T3-composion of composite, which is 3.49 GΩ (0.49×10^{11} Ohm.cm).

Table 4.4 shows resistance value (resistivity value) for five compositions of composites at a sintering temperature of 1000 °C.

The highest resistance value (resistivity value) is recorded for T3-composition of composite, which is 4.39 GΩ (0.62×10^{11} Ohm.cm).

Upon sintering at 1100 °C and 1200 °C, maximum resistance value (resistivity value) is exhibited again for T3-composion of composite. For T3-composion of composite sintered at 1100 °C and 1200 °C, resistance value (resistivity value) is found to be 7.07 GΩ (0.99×10^{11} Ohm.cm) and 132 GΩ (18.65×10^{11} Ohm.cm), respectively.

Table 4.3 Estimation of electrical resistivity of HCFC/kaolin composite sintered at 800 °C [radius=1.5 cm and thickness=0.5 cm].

S.no.	Sample ID	Resistance value (Gigaohm, GΩ)	Resistivity value (Ohm.cm)
1	T1	1.12	0.15×10^{11}
2	T2	1.55	0.21×10^{11}
3	T3	3.49	0.49×10^{11}
4	T4	0.97	0.13×10^{11}
5	T5	0.83	0.11×10^{11}

Note: T1 = HCFC slag (50%)/Kaolin (50%) composite; T2 = HCFC slag (60%)/Kaolin (40%) composite; T3 = HCFC slag (70%)/Kaolin (30%) composite; T4 = HCFC slag (80%)/Kaolin (20%) composite; T5 = HCFC slag (90%)/Kaolin (10%) composite.

Table 4.4 Estimation of electrical resistivity of HCFC/kaolin composite sintered at 1000 °C [radius=1.5 cm and thickness=0.5 cm].

S.no.	Sample ID	Resistance value (Gigaohm, GΩ)	Resistivity value (Ohm.cm)
1	T1	2.5	0.35×10^{11}
2	T2	3.2	0.45×10^{11}
3	T3	4.39	0.62×10^{11}
4	T4	1.49	0.21×10^{11}
5	T5	1.09	0.15×10^{11}

Note: T1 = HCFC slag (50%)/Kaolin (50%) composite; T2 = HCFC slag (60%)/Kaolin (40%) composite; T3 = HCFC slag (70%)/Kaolin (30%) composite; T4 = HCFC slag (80%)/Kaolin (20%) composite, T5 = HCFC slag (90%)/Kaolin (10%) composite.

Table 4.5 Estimation of electrical resistivity of HCFC/kaolin composite sintered at 1100 °C [radius=1.5 cm and thickness=0.5 cm].

S.no.	Sample ID	Resistance value (Gigaohm, GΩ)	Resistivity value (Ohm.cm)
1	T1	4.41	0.62×10^{11}
2	T2	5.2	0.73×10^{11}
3	T3	7.07	0.99×10^{11}
4	T4	6.43	0.90×10^{11}
5	T5	5.73	0.80×10^{11}

Note: T1 = HCFC slag (50%)/Kaolin (50%) composite; T2 = HCFC slag (60%)/Kaolin (40%) composite; T3 = HCFC slag (70%)/Kaolin (30%) composite; T4 = HCFC slag (80%)/Kaolin (20%) composite; T5 = HCFC slag (90%)/Kaolin (10%) composite.

It is therefore concluded that a composite having composition of 70% HCFC slag and 30% Kaolin sintered at 1200 °C has yielded the highest maximum resistance value (resistivity value) of 132 gigaohms (18.65×10^{11} Ohm.cm). This is attributed to the formation of mullite/spinel/oxides of chromium-bearing phase in the composite [30–32].

Hardness measured for T3 composition sintered at 1200 °C is shown in Table 4.7. The measurement has been taken by nanoindentation technique.

Table 4.6 Estimation of electrical resistivity of HCFC/kaolin composite sintered at 1200 °C [radius=1.5 cm and thickness=0.5 cm].

S.no.	Sample ID	Resistance value (Gigaohm, GΩ)	Resistivity value (Ohm.cm)
1	T1	44.1	**6.23** $\times 10^{11}$
2	T2	53.1	**7.50** $\times 10^{11}$
3	T3	132	**18.65** $\times 10^{11}$
4	T4	7.4	**1.04** $\times 10^{11}$
5	T5	1.42	**0.2** $\times 10^{11}$

Note: T1 = HCFC slag (50%)/Kaolin (50%) composite; T2 =- HCFC slag (60%)/Kaolin (40%) composite; T3 = HCFC slag (70%)/Kaolin (30%) composite; T4 = HCFC slag (80%)/Kaolin (20%) composite; T5 = HCFC slag (90%)/Kaolin (10%) composite.

Table 4.7 Hardness value of T3 composite.

Sample ID	$F_{max.}$ (mN)	HIT (MPa)	HVIT	EIT (MPa)	E* (GPa)
T3	99.82	9347.9	865.72	121.92	133.98

Note: T3 = HCFC slag (70%)/Kaolin (30%) composite.

The HVIT value estimated through nanoindentation is 865.72. The high Vickers value is due to formation of ceramics containing glassy phase and other strong crystalline phase. During sintering, there has been ordered bonding between atoms present in the components of composite [33].

An optical micrograph of T3 composite sintered at 1200 °C is shown in Figure 4.3, in which white patches of crystals that have spread in the matrix can be seen. Because of the poor contrast of the optical micrograph, phases are not clearly revealed. Therefore, gold-coated samples are examined under SEM at higher magnification.

Figure 4.4a,b shows two scanning micrographs of T3 composite sintered at 1200 °C at two different magnifications. Figure 4.4b is a magnifying micrograph of the same sample. At lower magnification, the sintered composite shows facets indicating brittleness of the material. There are some particles present in the micrograph. At a higher magnification (Figure 4.4b), smooth cleavage facets are revealed with some river pattern seen in the top corner of Figure 4.4b. There are some particles with different aspect ratios spread over the matrix. Thus, it is concluded that sintered material

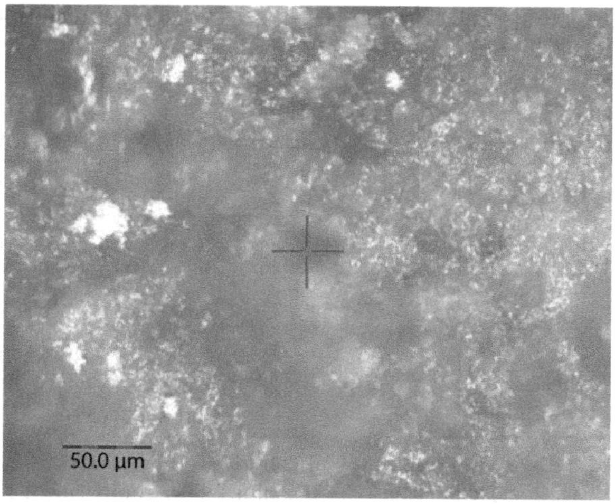

Figure 4.3 Optical image of HCFC slag (70%)/Kaolin (30%) composite sintered at 1200 °C.

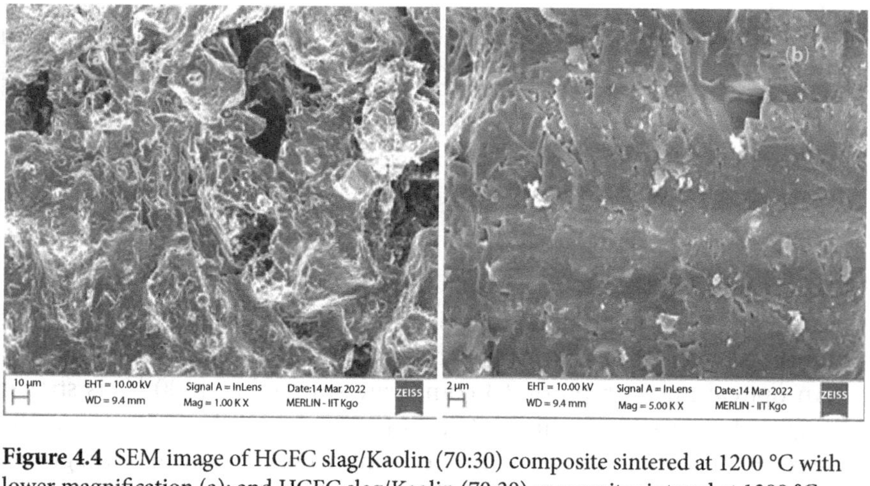

Figure 4.4 SEM image of HCFC slag/Kaolin (70:30) composite sintered at 1200 °C with lower magnification (a); and HCFC slag/Kaolin (70:30) composite sintered at 1200 °C with higher magnification (b).

possesses glassy phase as well as crystalline phases. There are some voids seen in the microstructure.

Along with SEM micrography, EDS analysis is conducted (Figure 4.5). The EDS spectrum (Figure 4.5) shows elements, i.e., O, Al, Si, Cr, Ca, Mg, Ti, Fe, Na and K. The EDS results are well-supported by XRD results.

Sintered composite is further examined by transmission electron microscopy in order to obtain a TEM micrograph of T3 composite sintered

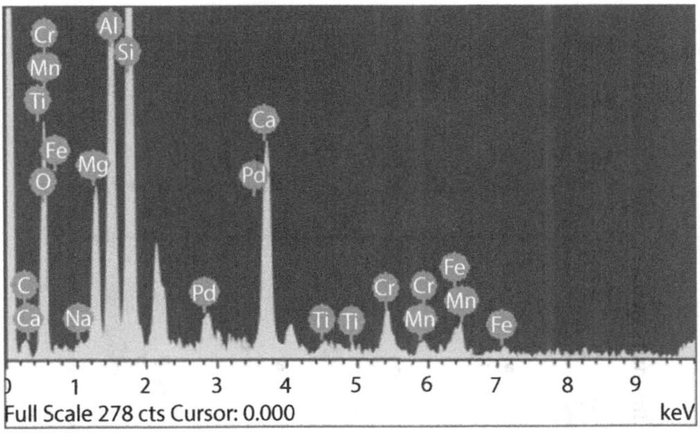

Figure 4.5 EDS spectrum of sintered (at 1200 °C) HCFC slag/Kaolin (70:30) composite.

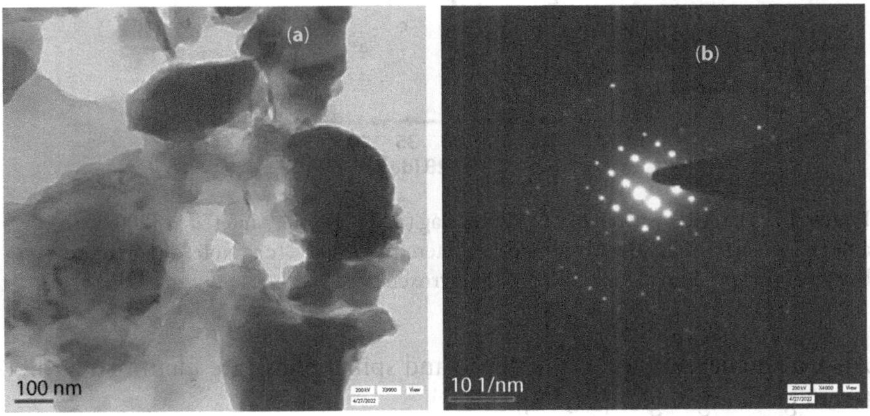

Figure 4.6 (a) TEM image of HCFC slag/Kaolin (70:30) composite sintered at 1200 °C and (b) SAD pattern of sintered HCFC slag/Kaolin (70:30) composite sintered at 1200 °C.

at 1200 °C. The purpose of the TEM study is to observe a micrograph with higher magnification and higher resolution.

Figure 4.6a shows the TEM image of T3 composite. Some fused crystals are dispersed in the matrix, which are crystalline in nature.

The X-ray diffraction pattern of T3 composite is shown in Figure 4.6b. Here, bright spots are concentrated with fewer spots present in a line. It is well-supported by the crystallinity of sintered samples.

Figure 4.7a shows the XRD pattern of raw HCFC slag. In the XRD pattern, various phases (i.e., forsterite, spinel (magnesium aluminate), and clinopyroxene) are recognized. The highest intensity phase is clinopyroxene.

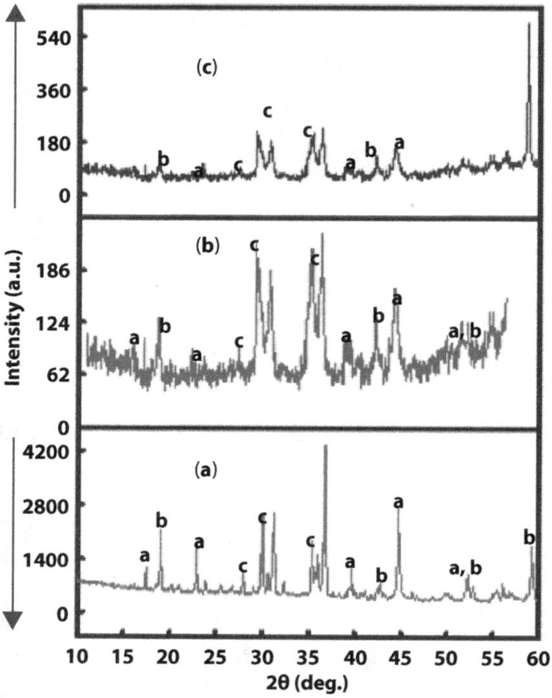

Figure 4.7 The XRD pattern of raw HCFC slag (a), HCFC slag sintered at 1200 °C (b), and HCFC slag/Kaolin (70:30) composite sintered at 1200 °C (c). [Note: a=Forsterite, b=Spinel (magnesium aluminate), c=Clinopyroxene]

Also, a smaller amount of forsterite and spinel phase are identified at different glancing angles [34, 35].

Figure 4.7b,c show the XRD pattern obtained for HCFC slag sintered at 1200 °C and HCFC slag/kaolin (70:30) composite sintered at 1200 °C (T3 composite). A distinct difference can be observed between unsintered and sintered product. In all cases, three major phases are identified (i.e., clinopyroxene, forsterite and spinel phase).

Intensity peaks and line-breath of unsintered sample are much different than that of sintered products. In unsintered slag sample, intensity is higher and line-breath is narrow. This indicates that slag material contains crystalline phases. While slag is sintered, peak intensity is dropped and line-breath is increased. This is due to sintering of slag at higher temperature, where interaction occurred between different crystalline phases present in the slag materials. For composite material, the number of peaks has decreased. Intensity or peak heights are reduced substantially, which indicates formation of a greater amount of non-crystalline phases, i.e., glassy phases. This is due to interaction between components present in kaolin and constituents present in HCFC slag.

The FTIR absorption bands of raw HCFC slag are shown in Figure 4.8a. Presence of different absorption stretching frequency of raw HCFC slag powder corresponds to raw HCFC slag components. The stretching frequency at 574 cm^{-1} matches with Si-O-Al band. Stretching frequencies are found at 1069 and 1602 cm^{-1}. This may be accredited to Si-O-Si asymmetric band and H-O-H bending band respectively. The Fe-O stretching band is also observed at 453 cm^{-1}; Al^{3+}-O^{2-} stretching band is also observed at 808 cm^{-1}; Si-O stretching band is also observed at 867 cm^{-1}; Cr^{3+}-O^{2-} stretching band is also observed at 1392 cm^{-1}, and Mg-O stretching band is also observed at 512 cm^{-1} [36–43].

The FTIR spectrum of HCFC slag sintered at 1200 °C is shown in Figure 4.8b, where broad stretching bands are found around 3451 cm^{-1} and

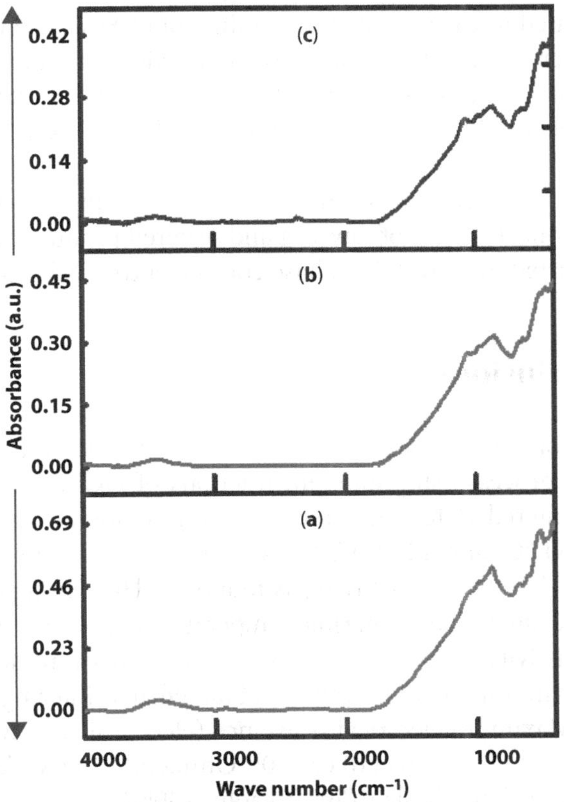

Figure 4.8 **The** FTIR spectra of raw HCFC slag (a), HCFC slag sintered at 1200 °C (b), and HCFC slag/kaolin (70:30) composite sintered at 1200 °C (c).

2941 cm^{-1}. The band arises due to the stretching vibrations of O-H bonds and H bending vibrations of H-O-H of interlayer adsorbed H_2O molecule [36-43]. C=O stretching band is indicated at 2248 cm^{-1} and confirms the presence of carbonate groups [36–43]. The main reason for this being the presences of chemisorbed CO_2 in HCFC slag sintered at 1200 °C. Si-O and O-Si-O bands are also found at 1100 cm^{-1} and 618 cm^{-1}, which indicates the presence of silicate groups. The presence of Al^{3+}-O^{2-} bonds is also observed near 808 cm^{-1}; and Fe-O stretching band is also observed at 468 cm^{-1}. The presence of Cr^{3+}-O^{2-} bond is also observed near 1178 cm^{-1}; and Mg-O bond is also observed near 808 cm^{-1} [36–43].

The FTIR spectrum of HCFC slag/kaolin (70:30) composite sintered at 1200 °C is shown in Figure 4.8c. Peak at 1092 cm^{-1} is attributed to Si-O band in SiO_4 molecules. A weak Si-O symmetrically stretching band is observed at 610 cm^{-1}. Peaks at 891 cm^{-1} and 702 cm^{-1} are assigned as alternating Si-O and Al-O bonds. Al-O stretching vibration is present in AlO_4 tetrahedra. There is shifting of stretching frequencies. It is observed that the band related to the asymmetric bending of O-Si-O and O-Al-O band shifts to lower value, which confirms the previous finding [36–43]. The Fe-O stretching band is also observed at 453 cm^{-1}; Cr^{3+}-O^{2-} stretching band is also observed at 1107 cm^{-1}; and Mg-O stretching band is also observed at 653 cm^{-1}.

From the FTIR results, it is concluded that stretching frequencies correspond to chemical groups of raw slag and sintered product constituents. It is well-supported in identifying phase component(s) (XRD results).

4.3 Conclusions

Composites of HCFC slag/kaolin ceramic were successfully produced via sintering process using different fractions of HCFC slag and kaolin. They were sintered at four different sintering temperatures, i.e., 800 °C, 1000 °C, 1100 °C, and 1200 °C. It was observed that maximum electrical resistance (electrical resistivity) is found for HCFC slag (70%)/kaolin (30%) composite for all sintering temperatures. For this composition, electrical resistivity value has increased with increase in sintering temperature. Maximum resistivity value is obtained at a sintering temperature of 1200 °C. Maximum electrical resistance (electrical resistivity) is determined to be 132 Gigaohms (18.65×10^{11} Ohm.cm). The Vickers hardness value of sintered HCFC slag (70%)/kaolin (30%) composite was determined to be 865.72.

The optical micrograph of sintered HCFC slag (70%)/kaolin (30%) composite displayed irregular sized particles. The SEM microstructure of sintered HCFC slag (70%)/kaolin (30%) composite showed fused mass. The TEM image of sintered HCFC slag (70%)/kaolin (30%) composite contained fused mass product with some crystalline phases, which was confirmed by SAD pattern. The EDS spectrum for sintered HCFC slag (70%)/kaolin (30%) composite shows the presence of O, Si, Al, Cr, Mg, Ca, etc., elements.

Mullite phase, spinel phases, forsterite phase, chromium-based phases, and clinopyroxene phase were identified in the XRD pattern.

The FTIR spectra detected chemical groups of peaks and described the formation of new bonds.

Acknowledgments

The authors would like to thank Prof. Munesh Chandra Adhikary, PG Council Chairman, Fokir Mohan University, for his invaluable guidance, advice, and constant inspiration throughout the entire process. They also would like to thank Mr. Mukteswar Mohapatra of Fokir Mohan University for his support. The authors convey their sincere thanks to GIET, University Gunupur, Rayagada, Odisha, India for providing lab facilities to do the research work. They would also like to thank CRF IIT Kharagpur for providing testing facilities.

References

1. A.I. Fares, K. Sohel, K. Al-Jabri, and A. Al-Mamun, (2021), Characteristics of Ferrochrome Slag Aggregate and Its Uses as a Green Material in Concrete-A Review, Construction and Building Materials,Vol. 294, pp. 123552-xxx.
2. P.H. Kumar, A. Srivastava, V. Kumar, and V.K Singh, (2014), Implementation of Industrial Waste Ferrochrome Slag in Conventional and Low Cement Castables: Effect of Calcined Alumina, Journal of Asian Ceramic Society, Vol. 2, pp. 371–379.
3. C. Kurtulus, R. Kurtulus, and T. Kavas, (2021), Foam Glass Derived from Ferrochrome Slag and Waste Container Glass: Synthesis and Extensive Characterizations. Ceramic International, Vol. 47, pp. 24997–25008.
4. M.Z. Islam, K.M.A. Sohel, K. Al-Jabri, and A. Al Harthy, (2021), Properties of Concrete with Ferrochrome Slag as a Fine Aggregate at Elevated Temperatures, Case Study of Construction Materials, Vol. 15, pp. e00599.

5. M. Erdem, H.S. Altundo˘gan, M.D. Turan, and F. Tümen, (2005), Hexavalent Chromium Removal by Ferrochromium Slag. Journal of Hazardous Materials, Vol. 126, pp. 176–182.
6. B.M. Moshtaghioun, and A. Monshi, (2007), Hot Corrosion Mechanism of Tundish Plaster with Steel Slags in Continuous Casting, Journal of Materials Science, Vol. 42, pp. 6720–6728.
7. L. Zhao, Y. Li, L. Zhang, and D. Cang, (2016), Effects of CaO and Fe_2O_3 on the Microstructure and Mechanical Properties of SiO_2–CaO–MgO–Fe_2O_3 Ceramics from Steel Slag, ISIJ International, Vol. 57, pp. 4–22.
8. D. Pei, Y. Li, S. Hua, S. Li, F. Jiang, and J. Yao, (2021), *In Situ* XRD Study on Function Mechanism of Pyroxene and Anorthite in Si-Ca Ceramics from Ferronickel Slag, Materials Letter, Vol. 305, pp. 130839-xxx.
9. D. Pei, Y. Li, and D. Cang, (2019), *In Situ* XRD Study on Sintering Mechanism of SiO_2-Al_2O_3-CaO-MgO Ceramics from Red Mud, Materials Letter, Vol. 240, pp. 229–232.
10. F. Jiang, Y. Li, L. Zhao, and D. Cang, (2017), Novel Ceramics Prepared from Inferior Clay Rich in CaO and Fe2O3: Properties, Crystalline Phases Evolution and Densification Process, Applied Clay Science, Vol. 143, pp. 199–204.
11. D. Pei, Y. Li, and D. Cang, (2017), Na+-Solidification Behavior of SiO_2-Al_2O_3-CaO-MgO (10 wt.%) Ceramics Prepared from Red Mud, Ceramic International, Vol. 43, pp. 16936–16942.
12. J. Burja, F. Tehovnik, J. Medved, M. Godec, and M. Knap, (2014), Chromite Spinel Formation in Steelmaking Slags, Materials and Tehnology, Vol. 48, pp. 753–756.
13. S. Barella, A. Gruttadauria, F. Magni, C. Mapelli, and D. Mombelli, (2012), Survey about Safe and Reliable Use of EAF Slag, ISIJ International, Vol 52, pp. 2295–2302.
14. G. Van Weert and M. Boering, (1995), Selective Pressure Leaching of Zinc and Manganese from Natural and Man-Made Spinels Using Nitric Acid, Hydrometallurgy, Vol. 39, pp. 201–213.
15. F. Bergaya and G Lagaly, (2006), General Introduction: Clays, Clay Minerals and lay science, Handbook of clay science, Vol. 1, pp. 1-8 (doi:10.1016/s1572-4352(05)01001-9).
16. C. N. Djangang, A. Elimbi, U. C. Melo, G. Lecomte-Nana, C. Nkoumbou, J. Soro, J. P. Bonnet, P. Blanchart, and D. Njopwouo, (2008), Sintering of Clay-Chamote Ceramic Composites for Refractory Bricks, Ceramics International, Vol. 34, pp. 1207-1213 (DOI: 10.1016/j.ceramint.2007.02.012).
17. J. (Theo) Kloprogge, Spectroscopic Methods in the Study of Kaolin Minerals and Their Modifications, Springer Mineralogy, Hardcover ISBN978-3-030-02371-3, eBook ISBN978-3-030-02373-7, pp.1-428 (DOI https://doi.org/10.1007/978-3-030-02373-7).
18. S. Mirmohamadsadeghi, T. Kaghazchi, M. Soleimani, and N. Asasian, (2012), An Efficient Method for Clay Modification and Its Application for

Phenol Removal from Wastewater, Applied Clay Science, Vol. 59-60, pp. 8-12 (https://doi.org/10.1016/j.clay.2012.02.016).
19. Blanca Bauluz Lázaro, (2015), Halloysite and Kaolinite: Two Clay Minerals with Geological and Technological Importance, Review of Real Academia de Ciencias. Zaragoza. Vol. 70, pp. 1-33.
20. S. K. Hubadillah, Z. Harun, Mohd H. D. Othman, and A. F. Ismail, and P. Gani, (2016), Effect of Kaolin Particle Size and Loading on the Characteristics of Kaolin Ceramic Support Prepared Via Phase Inversion Technique, Journal of Asian Ceramic Societies, Vol. 4, pp. xxx-xxx (DOI: 10.1016/j.jascer.2016.02.002).
21. J. C. Egbai, (2013), Kaolin Quantification in Ukwu-nzu and Ubulu-uku using Electrical Resistivity Method, International Journal of Recent Research and Applied Studies, Vol. 14, pp. xxx-xxx.
22. O. A. Babalola and T. Akomolafe, (2008), Effects of Kaolin Particle Size and Annealing Temperature on the Resistivity of Zinc-Kaolin Composite Resistors, Journal Applied Science Environmental Management, Vol. 12, p. 113-xxx.
23. L. Li and Y. Li, (2017), Development and Trend of Ceramic Cutting Tools from the Perspective of Mechanical Processing, IOP Conf. Series: Earth and Environmental Science, Vol. 94, pp.1-6.
24. M. Zhu, R. Ji, Z. Li, H. Wang, L. L. Liu, and Z. Zhang, (2016), Preparation of Glass Ceramic Foams for Thermal Insulation Applications from Coal Fly Ash and Waste Glass, Construction and Building Materials, Vol. 112, , pp. 398-405.
25. H. E. Exner and E. Arzt, (1996), Sintering Processes, In Physical Metallurgy, Ed. R. W. Cahn and P. Haasen, 4th Ed. Elsevier Science, Amsterdam, Vol.3, pp. 2628-2662.
26. W. V. Siemens, Inventor and Entrepreneur: Recollections of Werner Von Siemens, London, England, 1966.
27. T. Tanaka and T. Imai, (2013), Advances in Nanodielectric Materials Over the Past 50 years, IEEE Electrical Insulation Magazine, Vol. 1, pp. 10-23.
28. T. Tanaka, (2005), Dielectric Nanocomposites with Insulating Properties, IEEE Transactions on Dielectrics and Electrical Insulation, Vol. 12, pp. 914-928.
29. M. Panigrahi, R. I. Ganguly, R. R. Dash, (2018), An Advanced Cured High Carbon Ferrochrome Slag (HCFCS) Geopolymer (GP): A Constructional Materials in IOP Conference Series: Materials Science and Engineering, Vol. 410, pp.1-15.
30. M. K. Panigrahi, (2021), Investigation of Structural, Morphological, Resistivity of Novel Electrical Insulator: Industrial Wastes, Bulletin of Scientific Research, Vol. 3, pp. 51-58.
31. M. K. Panigrahi, (2022), Investigation of Structures of Sintered Fly Ash Materials: Resources of Industrial Wastes, Bulletin of Scientific Research, Vol. 4, pp.1-10.

32. Z. Zhan, W. Sun, S. Zhang, Q. Wen, H. Zhang, and X. Xiong, (2022), $(AlCrMg)_x(PO_4)_y/MgO$ composite: A New Thermal Protection and Insulation Material up to 2400 °C, Composites Part B: Engineering, Vol. 245, pp. 110198 (https://doi.org/10.1016/j.compositesb.2022.110198).
33. https://ceramics.org/about/what-are-engineered-ceramics-and-glass/structure-and-properties-of-ceramics
34. G. M. Biggar, and M. J. O'hara, (2006), Melting of Forsterite, Monticellite, Merwinite, Spinel, and Periclase Assemblages, Journal of the American Ceramic Society, Vol. 53, pp. 534 – 537 (DOI: 10.1111/j.1151-2916.1970.tb15959.x)
35. K.M.A. Sohel, Ali Fares, Khalifa S. Al-Jabri and Md. Abdullah Al-Mamun, (2021), haracteristics of ferrochrome slag aggregate and its uses as a green material in concrete – A review, Construction and Building Materials 294(2 August 2021):123552 (DOI: 10.1016/j.conbuildmat.2021.123552)
36. B.J. Saikia, G.K Rao, and P. Sarathy, (2010), Fourier Transform Infrared Spectroscopic Characterization of Kaolinite from Assam and Meghalaya, North Eastern India, International Journal of Modern Physcs, Vol. 1, pp. 206-xxx.
37. R.P. dos Santos, J. Martins, C. Gadelha, B. Cavada, A.V. Albertini, F. Arruda, M. Vasconcelos, E. Teixeira, F. Alves, J.L. Filho, and V. Freire, (2014), Coal Fly Ash Ceramics: Preparation, Characterization, and Use in the Hydrolysis of Sucrose. Scientific World Journal, Vol. 2014, pp. 01.
38. A. Jose, M. R. Nivitha, M. Krishnan, and R. G. Robinson, (2020), Characterization of Cement Stabilized Pond Ash using FTIR Spectroscopy, Construction and Building Materials, Vol. 263, pp.120136-xxx (DOI: 10.1016/j.conbuildmat.2020.120136).
39. Y. Liu, F. Zeng, B. Sun, P. Jia, and I. T. Graham, (2019), Structural Characterizations of Aluminosilicates in Two Types of Fly Ash Samples from Shanxi Province, North China, Minerals, Vol. 9, pp. 358-16 (doi:10.3390/min9060358).
40. P. Rożek, M. Król, and W. Mozgawa, (2018), Spectroscopic Studies of Fly Ash-Based Geopolymers, Spectrochim Acta - Part A Molecule, Biomolecule Spectroscopy Vol. 198, pp. 283-289 (https://doi.org/10.1016/j.saa.2018.03.034).
41. S. Kumar, F. Kristály, and G. Mucsi, (2015), Geopolymerisation Behaviour of Size Fractioned Fly Ash, Advanced Powder Technology, Vol. 26, pp. 24-30 (https://doi.org/10.1016/j.apt.2014.09.001).
42. Z. Kledyński, A. Machowska, B. Pacewska and I. Wilińska, (2017), Investigation of Hydration Products of Fly Ash-Slag Pastes, Journal of Thermal Analysis and Calorimetry, Vol. 130, pp. 351-363 (https://doi.org/10.1007/s10973-017-6233-4).
43. D. Bondar and R. Vinai, (2022), Chemical and Microstructural Properties of Fly Ash and Fly Ash/Slag Activated by Waste Glass-Derived Sodium Silicate, Crystals, Vol. 12, pp. 913-12 (https:// doi.org/10.3390/cryst12070913).

5

High Resistant HCFC Slag/ K-Feldspar Ceramics

Muktikanta Panigrahi[1]*, Ratan Indu Ganguly[2] and Radha Raman Dash[3]

[1]*Department Materials Science, Maharaja Sriram Chandra Bhanja Deo University, Keonjhar Campus, Odisha, India*
[2]*Department of Metallurgical Engineering, National Institute of Technology, Rourkela, Odisha, India*
[3]*CSIR-National Metallurgical Laboratory (NML), Jamshedpur, Jharkhand, India*

Abstract

During production of ferrochrome alloy, a large amount of slag is produced as a waste product. Dumping of these waste materials pollutes the environment. Hence, it is deemed necessary to utilize this waste to develop a useful material. This will also provide socioeconomic benefits.

Experiments have been conducted to develop HCFC slag/K-feldspar ceramic composite. For this, an appropriate amount of raw components, i.e., HCFC slag, K-feldspar and dextrin, are mixed uniformly and compacted in a mold applying pressure (10 MPa) with a UTM machine. Demolded samples are sintered at a sintering temperature of 1200 °C (for 1 h). A comprehensive investigation report is prepared. Analyses of structure co-relating with properties are described; and materials are characterized using sophisticated instruments.

Electrical resistance value (resistivity value) of the sintered composite has been estimated by two-probe technique. Maximum electrical resistance value (resistivity value) is found to be 265.1 Gigaohms (37.47×10^{11} Ω.cm). Resistivity value indicates the possibility of using this material as a good insulator for engineering applications. This will be a cheaper product since the cost of raw material will be less. The Vickers hardness value of the sintered product is arrived at by nanoindentation and is measured to be 678.95. By XRD analyses, forsterite, spinel, quartz, mullite, and chromium-bearing phases are identified. The presence of mullite phase enhances the insulating properties of the sintered materials. Microstructural analyses of

**Corresponding author*: muktikanta2@gmail.com

Muktikanta Panigrahi, Ratan Indu Ganguly and Radha Raman Dash (eds.) High Electrical Resistant Materials: Ferrochrome Slag Resource Ceramics, (117–138) © 2024 Scrivener Publishing LLC

sintered sample are done by SEM and TEM. Presence of different chemical groups of samples is revealed by FTIR analyses.

Keywords: Industrial wastes, slag, XRD, phases, mullite, insulator, morphology, resistivity

5.1 Introduction

Ferrochromium industries generate large amounts of ferrochrome (FeCr) slag. Usually, for production of 1 ton of FeCr, the amount of slag generated is 1–1.2 t. In recent years, with the increasing demand for stainless steel, the requirement of FeCr alloy has increased. Consequently, the substantial volume of FeCr slag is increasing [1], with FeCr slags piling up in the open air. In prolonged storage in soil, FeCr slag can be leached and enter soil due to rainwater, thereby threatening the environment. Therefore, effective use of FeCr slags has become a traditional topic for research. Zelic [2], Panda *et al.* [3] and Gencel *et al.* [4] have reported preparation of concrete aggregates from FeCr slag. Zhang *et al.* [5] synthesized high-strength lightweight aggregates with FeCr slag. Gencel *et al.* [6] used FeCr slag to prepare bricks which have achieved mechanical strength greater than 7 MPa. Liu *et al.* [7] synthesized porous cordierite ceramics using FeCr slag as a raw material, which yielded a flexural strength of 47.26 MPa. Available reports indicate fruitful utilization of FeCr slag in the field of building materials. Furthermore, production of high-strength dense ceramics using FeCr slag is thought to be an important proposition for developing ceramic insulators.

Bauxite-based castable and cement castables have drawn attention for their effective use as liner in coke oven batteries, coal-fired power boilers, industry furnaces, iron industries, etc. [8]. Addition of micro silica in these castables improves refractory properties, and increases packing density [9–12]. Weakening of deposits of ore and non-ore minerals results in more ecological imbalance and increases cost of production. However, by effective use of these materials and secondary resources, it is possible to develop low-cost castables required for liners. Thus, secondary resources will minimize environmental problems [13].

In the production of refractories, there is an option of metallurgical industry wastes to be utilized. Slags formed in production of high-carbon ferrochrome are usually dumped. Utilization of dumped ferrochrome slag in refractory castables reduces the cost of product and is friendly to the atmosphere [14–18].

Global ferrochrome production is more than 8.9 million tons. The world's major ferrochrome producers are South Africa (42%), China (23%), Kazakhstan (13%), and India (10%) [19].

Ferrochromium is a master alloy of iron and chromium. During its production from chromite ore, carbon is picked-up by alloy, resulting in high-carbon ferrochromium alloy. A raw material used for production of FeCr is chromite ore ($FeO{\cdot}Cr_2O_3$). Chromite ore is a chrome oxide- and iron oxide-bearing mineral. Chromite ore is lumpy and is used after suitable sizing. Fine particles generated during procurement of ore are agglomerated. For efficient production of alloy, fluxing materials used are quartzite, bauxite, dolomite, corundum, lime and olivine. This enables the elimination of undesirable oxides and slags [20]. Major components of slag are SiO_2, MgO, and Al_2O_3. Other components of slag, such as Cr_2O_3, Fe_2O_3 and CaO, are also present. Common phases present in the slag are glass, spinels (Al_2O_3-MgO), forsterite (MgO-SiO_2) [21–23] and CaO.

For preparation of HCFC slag-based high-resistance ceramics, feldspar is mixed into the slag. Feldspar is used because of its availability in the Earth's upper crust (about 50 wt%) [24]. Feldspars are the main rock-forming minerals in magmatic, metamorphic, and sedimentary rocks. Hence, their conductivity is likely to have a large influence on the electrical conductivity of rocks in Earth's crust. Studies on the conductivity of alkali feldspar aid in the exploration of electrical properties of the entire feldspar family. Electrical conductivity measurements on synthetic as well as natural feldspar have been extensively performed in a laboratory [25]. However, most of these previous studies on electrical conductivity of feldspar have been carried out at ambient pressure and high temperature. Pressure-dependent glass transition temperature (T_g) has been studied [26]. Pressure has an effect on the activation energy for electrical conduction in albite, haplogranite, anorthite, and SiO_2 glasses. Ni *et al.* [27] have measured electrical conductivity of dry and hydrous albite glasses and liquids at high pressure and high temperature. For both samples, feldspar glasses were used, not the natural crystalline feldspar that commonly exists in the Earth's crust. Yang *et al.* [28] measured the electrical conductivity of dry and hydrous plagioclase at high temperature and pressure on both natural and pre-annealed samples obtained from fresh mafic granulite xenoliths. Alkali feldspar is not used in the samples as the influence of chemical composition on electrical conductivity of feldspar has not been quantitatively evaluated. An important finding from previous studies is that the conduction mechanism in feldspar is described as due to migration of alkali cations (Na^+, K^+ and Ca^{2+}). It is not yet understood whether alkali cations transport by a vacancy mechanism or by an interstitial mechanism. Thus, controversy exists regarding alkali ions dynamics in the diffusion mechanism occurring in feldspar. The electrical conductivity of albite has been thoroughly studied in previous work by Hu *et al.* [29]. However, they have not completely solved the above problems. Consequently, it is very important to clarify the dynamics of charge transport in alkali feldspar under high temperature. A systemic study of electrical conductivity of alkali feldspar with different composition at high temperature and high pressure is very important.

Current investigations are being conducted to develop a ceramic insulator from ferrochrome slag, as such ceramics have many applications, including in refractories [30], cutting tools [31], thermal insulator [32], etc. Processing of ceramic material involves the procurement and processing of raw materials, and compaction of raw materials followed by firing/sintering at elevated temperature. During firing/sintering, there is interaction between ingredients present in raw materials and formation of various binary or ternary crystalline phases. Proper compaction and densification through sintering enhance properties. During sintering, ingredients are usually present in raw materials and formation of various binary or ternary crystalline phases. Therefore, proper processing/sintering needs to be ensured to obtain high-quality ceramics. Raw materials consisting of alumina-silica in proper proportion will enable the formation of mullite. Mullite phase formation in the system can be completed at 1150–1200 °C; therefore, high-temperature sintering is essential [33].

Electrical insulators are categorized according to insulation properties [34, 35]. In present scenarios, there is a growing demand for low-cost insulator in the area of electrical engineering. Development of such insulator has become a necessity, and therefore researchers are trying to develop cheaper insulator having the requisite properties [36].

Ceramics are extensively used in electronic components such as capacitors, transducers, etc. At the present time, fast-growing capacitor technologies have drawn attention due to their many applications. With advancement in microelectronics and communication engineering, the need for such materials has been realized in the technologies. It is envisaged that more applications of ceramic capacitors will be found in the future [37].

In the current work, sintered high-carbon ferrochrome (HCFC) slag-based materials are prepared through sintering process via solid-state route. Materials are prepared at different sintering temperature of 1200 °C. Sintered materials are analyzed by SEM and EDS. Different phases of present sintered materials have been identified by X-ray diffractometry; and FTIR analysis is done to determine the presence of different chemical groups in the sintered samples. Electrical resistivity value is also calculated.

5.2 Experiment Details

5.2.1 Materials and Chemicals

The HCFC slag is procured from Shyam Ferro Alloys, Durgapur, West Bengal, India. Composition of HCFC slag is shown in Table 5.1. K-feldspar is procured from Loba Chemie, along with Dextrin $(C_6H_{10}O_5)_n{\cdot}xH_2O)$, which is used to bind the HCFC slag particles via sintering reaction process. During HCFC slag ceramic material preparation, a few drops of water (i.e., 6%) is used. Optical images of HCFC slag, dextrin, and K-felspar are shown in Figure 5.1.

5.2.2 Materials Preparation

Preparation of HCFC slag-based sintered materials is achievable through solid-state route [40, 41]. Two types of sintered materials (i.e., sintered HCFC slag and sintered HCFC slag/K-feldspar) are prepared.

5.2.2.1 HCFC Slag-Sintered Materials

The following steps are adopted to obtain HCFC slag sintered material:

Step 1: Processing of HCFC slag
The HCFC slag is ground in a ball mill (5 h). It is sieved through 240 meshes. Sieved HCFC slags are dried in a heating oven at 120 °C (2 h). The purpose of heating sieved HCFC slag is to remove moisture. Such HCFC slag is called processed HCFC slag.

Step 2: Preparation of green pellet
Appropriate amount (10 g) of processed HCFC slag (i.e., 100% HCFC slag) is put into a mortar with pestle. A small amount of dextrin (i.e., 0.05%) is added to it as a binder. Both ingredients are ground in a mortar with a pestle (1 h). The purpose of grinding is to mix ingredients uniformly. A small amount of water is used to make the mixture pasty.

Pellet is made from the above paste. The paste is kept in a clean pelletizer. The pelletizing mold is kept under pressure with a universal testing machine (UTM). During compaction, pressure is maintained at 10 MPa for five minutes. Thereafter, pelletizer is removed from the molding system and pellets are collected by demolding the molding system. Dimensions (thickness and diameter) of pellet(s) are measured. It is estimated to be

Table 5.1 Compositions with percentage (%) of processed HCFC slag and feldspar (potassium).

S. ID	Compositions (%)												
	Fe_2O_3	Al_2O_3	SiO_2	FeO	Cr_2O_3	Na_2O	MgO	K_2O	CaO	TiO_2	BaO	LOI	Ref
HCFC slag	2.5-3.0	23.5-24.5	28-31	---	4.5-5.4	---	26-26.5	---	7.5-8.5	---	---	---	[38]
k-FD	---	18.18	64.57	---	0.22	---	---	---	---	0.01	----	---	[39]

Note: S. ID = Sample ID, HCFC slag = High-carbon ferrochrome slag, k-FD = Potassium Feldspar.

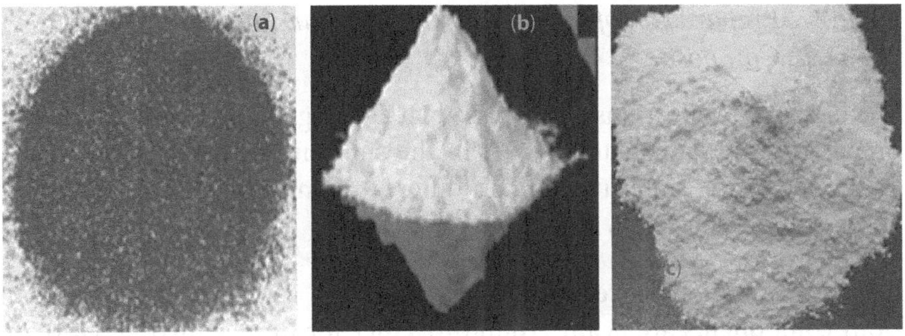

Figure 5.1 Optical images of HCFC slag (a), Dextrin (b) and K-feldspar (c).

1.1 mm (average thickness) and 10 mm (average diameter). Such pellet is known as green pellet.

Step 3: Preparation sintered HCFC slag materials
Green pellet samples are kept in a muffle furnace. They are bisquetted at 900 °C for 2 h to remove water and binder. Bisquetted pellet(s) are sintered at 1200 °C for 1 h. Average dimensions (i.e., thickness and diameter) of sintered pellets are measured to be 1.09 mm and 10 mm, respectively. The entire process is shown schematically in Figure 5.2. Sintered pellet(s) are now ready for different characterization processes using sophisticated instruments.

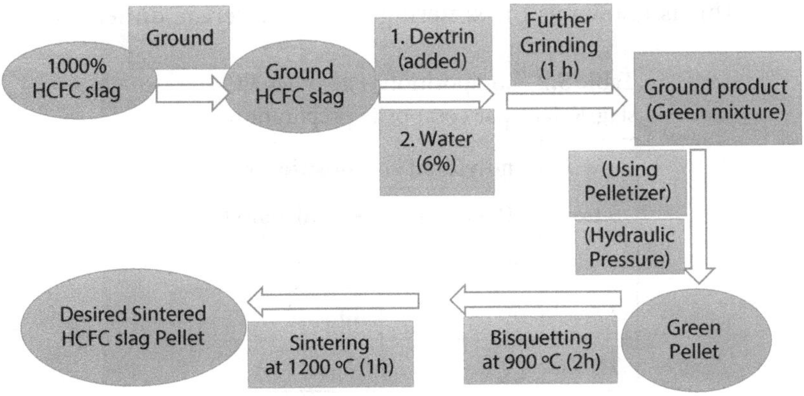

Figure 5.2 Flow chart showing the preparation of sintered HCFC slag materials.

5.2.2.2 *Preparation of HCFC Slag/K-Feldspar-Based Sintered Composite*

The different proportions of processed HCFC slag and K-feldspar used for preparation of composite [40, 41] in the sintering process are shown in Table 5.2. In the sintering process, different compositions of composites are prepared (Table 5.2).

Step 1: Preparation of green pellet
Both components are again ground together for proper mixing. Dextrin (0.05%) is used as a binder. A few drops of water are used to make the mixture pasty.

The mixed HCFC slag/K-feldspar paste is transferred to a clean pelletizer. It is then kept under pressure with a universal testing machine (UTM). Compaction pressure is kept at 10 MPa. It is left for five minutes for their proper compaction. Then, pelletizer is removed from molding system. Pellets are collected after demolding the molding system. Average thickness and average diameter of pellet(s) are measured. Average thickness is found to be 1.1 mm and average diameter is found to be 10 mm. Such pellet is known as green pellet.

Step 2: Sintering of green pellet
Green HCFC slag/ K-feldspar composite pellet is placed in a muffle furnace. Temperature of muffle furnace is maintained at 900 °C. At this temperature, pellet samples are bisquetted for 2 h. At this temperature, this will ensure removal of water and dextrin binders. This process is known as the bisquetting process. Following this process, pellet(s) are fired at 1200 °C for 1 h. This is known as the sintering process. Average dimensions (i.e.,

Table 5.2 Compositions (%) of HCFC slag/K-feldspar composite preparation.

S.no.	Individual composition name	
	HCFC slag	**K-feldspar (FD)**
1	50	50
2	**60**	**40**
3	70	30
4	80	20
5	90	10

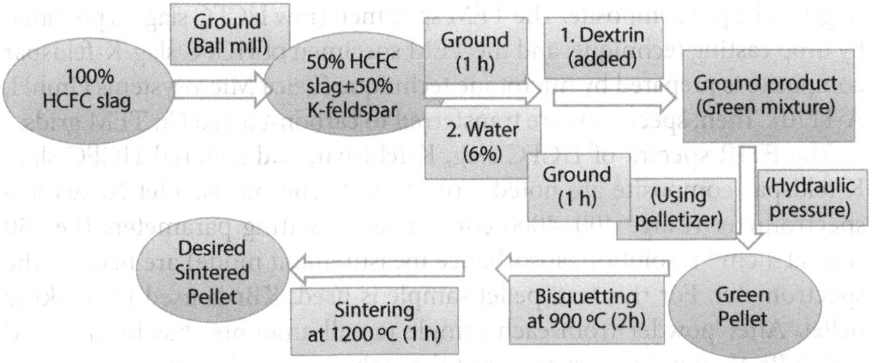

Figure 5.3 Flow chart showing the preparation of HCFC slag (50%)/K-feldspar (50%) composite sintered at 1200 °C.

average thickness and diameter) of sintered pellets are found to be 1.09 mm and 10 mm, respectively. The entire process is presented schematically in Figure 5.3.

The sintered pellet(s) is then ready for use in different tests.

Other sets (Table 5.2) of sintered HCFC slag/K-feldspar composite is prepared in a similar manner.

5.2.3 Test Methods

X-ray diffraction (XRD) analyses are used to identify different phases. For these analyses, a Philips PW-1710 Advance wide-angle X-ray diffractometer and Philips PW-1729 X-ray generator instrument are used. CuKα radiation of wavelength (λ) 0.154 nm is used for XRD analyses. The X-ray diffractometer functions at 40 kV and 20 mA. Prepared pellet sample is put on a sample holder (made from quartz). Measurement is performed at room temperature. Sample is scanned from 20° to 70° with scanning speed (2°/min).

An optical image is used for measurement of hardness by a nanoindentation tester (NHT; Anton Paar, Switzerland).

Morphological analyses are done with a Carl Zeiss Supra-40 scanning electron microscope (SEM). Sintered samples are gold-coated by sputtering technique. During testing, the instrument parameter, i.e., operating voltage, is kept at 30 kV. Along with SEM, EDS analysis is also carried out.

Also, sintered samples are examined by a transmission electron microscope (TEM; 5022/22, Tecnai G220 S-Twin, Czech Republic). TEM is used to investigate images and SAD pattern of raw HCFC slag and HCFC slag/K-feldspar composite. The TEM specimen (raw HCFC slag) is prepared by drop casting technique; and the TEM specimen of HCFC slag/K-feldspar composite is prepared by microtone technique (Leica Microsystems GmbH, A-1170). Then, specimens are transferred to carbon-coated Cu TEM grids.

The FTIR spectra of HCFC slag, K-feldspar, and sintered HCFC slag/K-feldspar composite are noted. For the test, Thermo Nicolet Nexus 870 spectrometer (range 400–4000 cm^{-1}) is used. Setting parameters (i.e., 50 scan at 4 cm^{-1} resolution, absorbance measurement mode) are fixed in the spectrometer. For this test, pellet sample is used. KBr is used for making pellet. After powder from each sample (small amounts) has been mixed with KBr, then pellets are prepared through compression molding system and pelletizer. A background spectrum is collected before testing. Then, the pellet is put in an FTIR sample holder and data are collected.

For electrical resistivity measurement, sintered pellet has been used. Resistance is measured for the material using the two-probe method. Resistivity is calculated by the following equation [40, 41]

$$\rho = \left(\frac{R \times A}{l}\right) \tag{5.1}$$

Here, A is cross-sectional area of electrode $= \frac{\pi}{4} D^2$ (m^2), R is the resistance of prepared samples (GΩ), l is thickness of the sample, ρ is the resistivity of the samples in Ω.cm. For this measurement, a multimeter is used.

Hardness measurement is taken by a nanoindentation tester (NHT; Anton Paar, Switzerland).

5.2.4 Results and Discussion

Table 5.3 shows the resistivity value of HCFC/K-feldspar composites sintered at 1200 °C. From the estimated resistivity data (Table 5.3), the highest resistivity value is found to be 37.47×10^{11} Ohm.cm for HCFC slag (70%)/K-feldspar (30%) composite; and the lowest resistivity value is found to be 0.35×10^{11} Ohm.cm for HCFC slag with dextrin. It is concluded that the highest resistivity value is due to formation of mullite phases, magnesium aluminate phase, and oxides of chromium-based complex phases [40–42].

Vickers hardness value of HCFC slag (70%)/K-feldspar (30%) sintered composite is presented in Table 5.4. The Vickers hardness value is found to be 678.95.

Table 5.3 Estimation of electrical resistivity of HCFC/K-feldspar sintered (1200 °C) composite [radius=1.5 cm and thickness=0.5 cm].

S.no.	Sample ID	Resistance value (Gigaohm, GΩ)	Resistivity value (Ohm.cm)
0	T0	2.46	0.35×10^{11}
1	T1	220	31.1×10^{11}
2	T2	250	35.34×10^{11}
3	T3	265.1	37.47×10^{11}
4	T4	72.34	10.23×10^{11}
5	T5	23.44	3.313×10^{11}

Note: T0 = HCFC slag with dextrin; T1 = HCFC slag (50%)/K-feldspar (50%) composite; T2 = HCFC slag (60%)/K-feldspar (40%) composite; T3 = HCFC slag (70%)/K-feldspar (30%) composite; T4 = HCFC slag (80%)/K-feldspar (20%) composite; T5 = HCFC slag (90%)/K-feldspar (10%) composite.

Table 5.4 Vickers hardness value of HCFC slag (60%)/K-feldspar (40%) sintered composite.

S.no	Sample ID	$F_{max.}$ (mN)	HIT (MPa)	HVIT	EIT (MPa)	E* (GPa)
1	T1	99.80	7331.3	678.95	129.64	142.46

Note: T1 = HCFC slag (70%)/K-feldspar (30%) composite.

An optical image of HCFC slag (70%)/K-feldspar (30%) composite sintered at 1200 °C is shown in Figure 5.4. It can be seen from the figure that white patches of crystals are spread in the matrix. There are irregular shaped particles having different textures, which are dispersed in the matrix. The optical micrograph shows poor contrast; therefore, an SEM study is carried out to understand these features in detail.

Figure 5.5 shows the SEM images of raw HCFC slag and raw K-feldspar. It can be seen from the SEM image of HCFC slag (Figure 5.5a) that HCFC slag contains irregular shaped crystals with regular smooth texture. Irregular shaped crystals are dispersed in the background [43]. It can be seen from the SEM image of K-feldspar (Figure 5.5b) that K-feldspar contains irregular shaped crystals with regular smooth texture and different aspect ratio. Irregular shaped crystals are dispersed in the background.

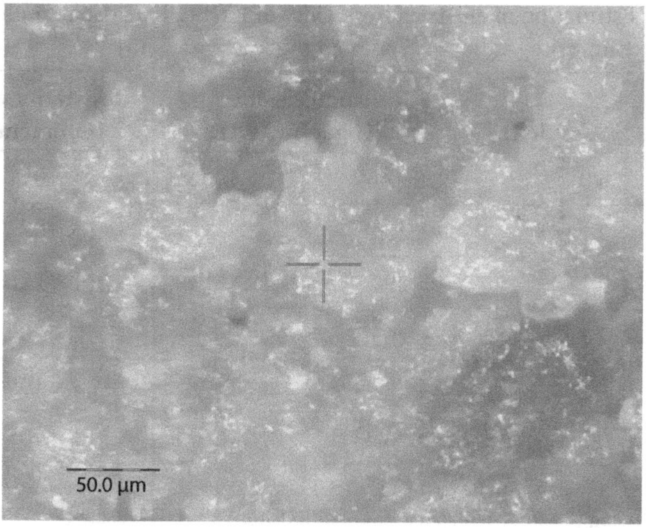

Figure 5.4 Optical image of HCFC slag (70%)/K-feldspar (30%) sintered composite.

Figure 5.5 SEM images of raw HCFC slag (a) and SEM image of raw K-feldspar (b) [43].

The SEM images of sintered HCFC slag and HCFC slag/K-feldspar (70:30) composite are shown in Figure 5.6a,b. Sintered HCFC slag contains glassy phase with smooth texture. A few areas contain some crystalline phases. There is a marked difference between sintered HCFC slag (Figure 5.6a) if compared with unsintered HCFC slag (Figure 5.5a). Figure 5.6b shows sintered HCFC slag/K-feldspar composite. The topology shows fused masses with pores in the lumpy fussed mass. The surface topology is

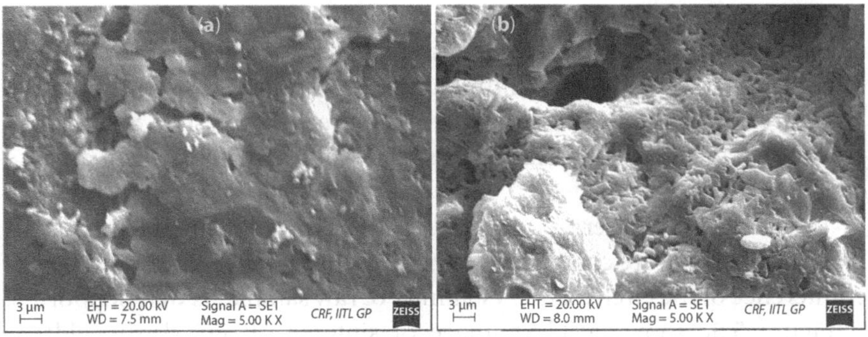

Figure 5.6 SEM image of HCFC slag sintered at 1200 °C (a), and SEM image of HCFC slag/K-feldspar (70:30) composite sintered at 1200 °C (b).

rough. Some crystalline phases are located in some spots. A marked difference is observed in the sintered micrograph if compared with raw HCFC slag and raw K-feldspar.

The EDS analyses are done together with SEM studies. Elements like O, Al, Si, Ti, Fe, Cr, Mg, Ca, Na and K are identified in all samples. The EDS results are well-corroborated with XRD results.

Figure 5.7a shows the TEM image of raw HCFC slag. As can be seen in Figure 5.7a, hexagonal shaped crystals are dispersed in the matrix, which are crystalline in nature.

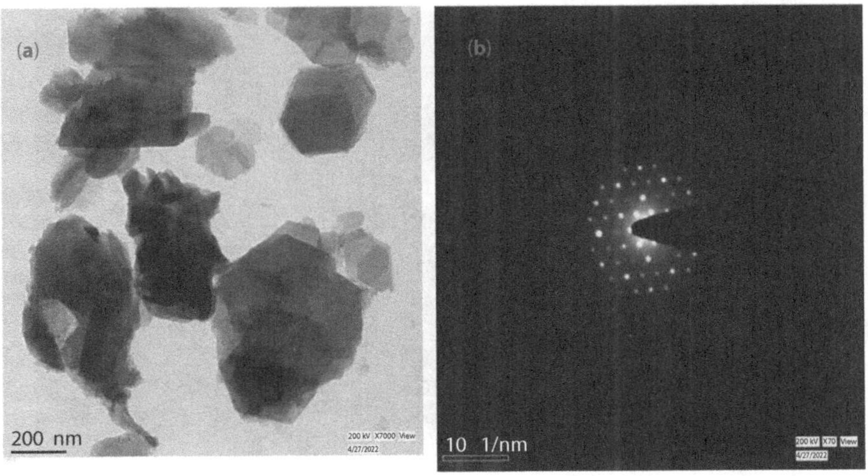

Figure 5.7 TEM images of raw HCFC slag (a), and SAD pattern of raw HCFC slag (b).

The diffraction pattern of raw HCFC slag is shown in Figure 5.7b. Here, bright spots are concentrated with lesser spots present around the circumference of the circle. The crystalline nature of raw samples is well-supported.

Figure 5.8a shows a TEM image of sintered HCFC slag. Some fused crystals of sintered HCFC slag are dispersed in the matrix, which are also crystalline in nature.

The diffraction pattern of sintered HCFC slag is shown in Figure 5.8b. Here, bright spots are concentrated with lesser spots spread around the circumference of the circle. The crystallinity of raw samples is well-supported.

Figure 5.8c shows a TEM image of sintered HCFC slag/K-feldspar (70:30) composite. Some fused crystals of sintered HCFC slag are dispersed in the matrix, which are also crystalline in nature.

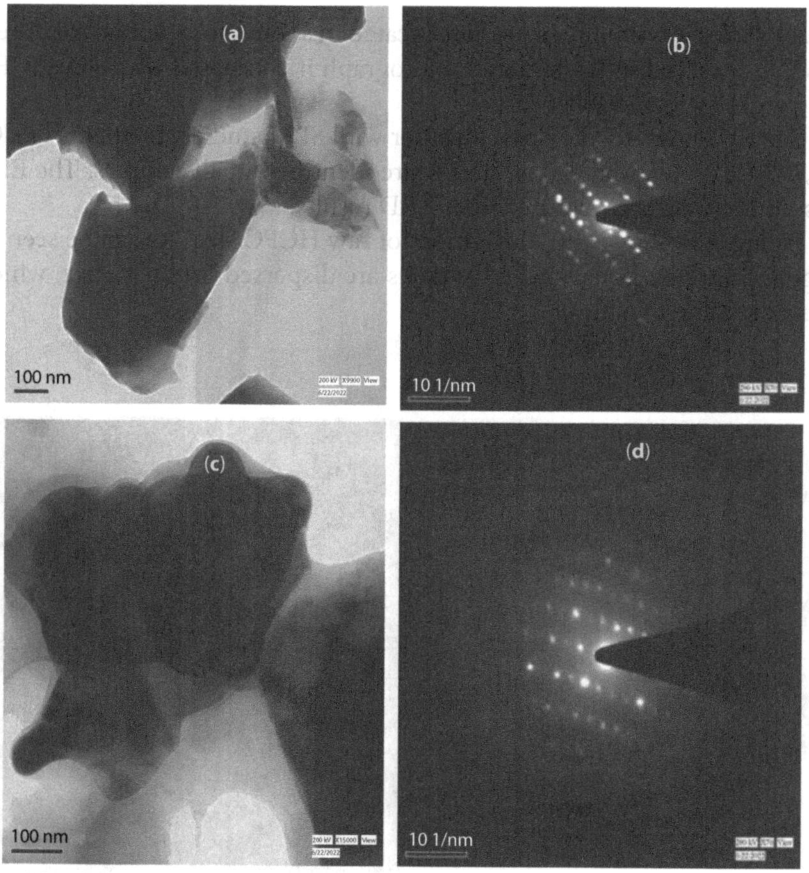

Figure 5.8 TEM image of HCFC slag sintered at 1200 °C (a); SAD pattern of sintered HCFC slag (b); TEM image of HCFC slag/K-feldspar (70:30) composite sintered at 1200 °C (c); and SAD pattern of sintered HCFC slag/K-feldspar (70:30) composite (d).

The diffraction pattern of sintered HCFC slag/K-feldspar (70:30) composite is shown in Figure 5.8d. Here, bright spots are concentrated with lesser spots spread around the circumference of the circle. The crystallinity of raw samples is well-supported.

Figure 5.9a shows the XRD pattern of raw HCFC slag. In the XRD pattern, various phases (i.e., forsterite ($MgSiO_4$), wollastonite ($CaSiO_3$), magnesium aluminate (spinel~$MgAl_2O_4$), chromium-bearing phases [($Na_{45}Ca_{55}$)($Cr_{45}Mg_{55}$)$52O_2$ and mullite) are recognized. The highest intense peak is found at 36° (2Θ) in the XRD pattern (Figure 5.9a). It corresponds to chromium-bearing phases [($Na_{45}Ca_{55}$)($Cr_{45}Mg_{55}$)$52O_2$.

Figure 5.9b shows the XRD pattern of HCFC slag, which is sintered at 1200 °C. Various peaks are observed in the pattern. The highest intense peak is found at 36° (2Θ) in the XRD pattern (Figure 5.9b) and the phase is silica (ICSD analysis). The other silica phase is observed at 42°, 74°, and

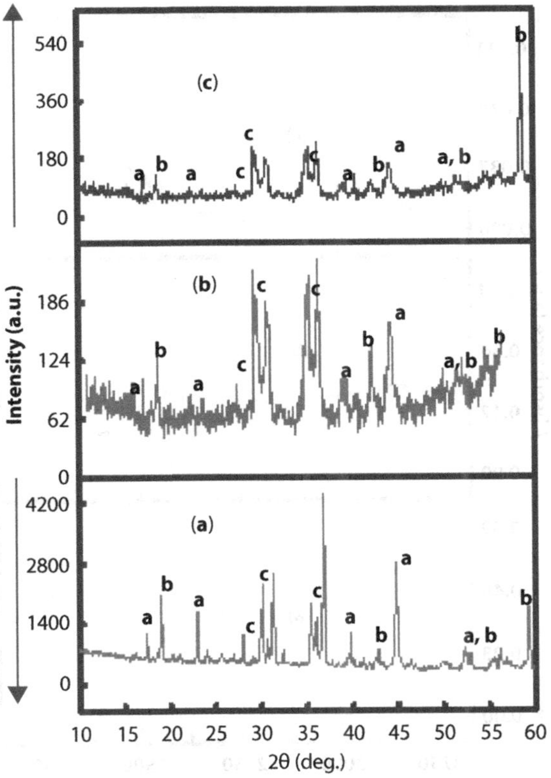

Figure 5.9 XRD pattern of HCFC slag (a); sintered HCFC slag sintered at 1200 °C (b); and HCFC slag/K-feldspar (70:30) composite sintered at 1200 °C (c). [Note: a=forsterite, b=spinel (magnesium aluminate), c=clinopyroxene].

79°, respectively. Other phases, such as mullite and alumina, may be present. Figure 5.9c shows the XRD pattern of HCFC slag/K-feldspar (70:30) composite sintered at 1200 °C. Different peaks, such as forsterite ($MgSiO_4$), wollastonite ($CaSiO_3$), magnesium aluminate (spinel~$MgAl_2O_4$), chromium-bearing phases $(Na_{45}Ca_{55})(Cr_{45}Mg_{55})52O_2$ and mullite are observed in the pattern. The highest intense peak is found at 36° (2Θ) in the XRD pattern (Figure 5.9c) and the phase is silica (ICSD analysis). The silica phase belongs in hexagonal crystal system.

The FTIR absorption bands of raw HCFC slag are shown in Figure 5.10a. Presence of different absorption stretching frequency of raw HCFC slag powder corresponds to raw HCFC slag components. The stretching frequency at 574 cm^{-1} matches with Si-O-Al band. Stretching frequencies are found at 1069 and 1602 cm^{-1}. This may be accredited to Si-O-Si asymmetric band and H-O-H bending band respectively. The Fe-O stretching

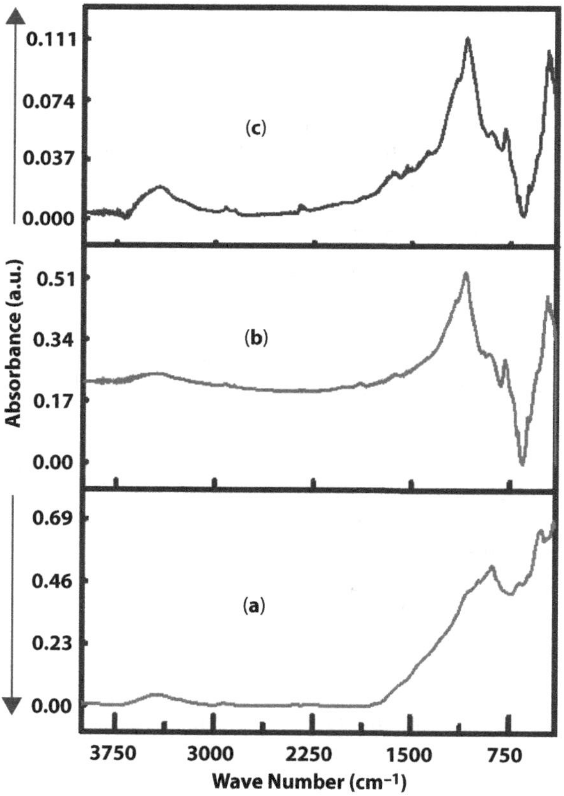

Figure 5.10 FTIR spectrum of HCFC slag (a); sintered HCFC slag sintered at 1200 °C (b); and HCFC slag/K-feldspar (70:30) composite sintered at 1200 °C (c).

band is also observed at 453 cm^{-1}; Al^{3+}-O^{2-} stretching band is also observed at 808 cm^{-1}; Si-O stretching band is also observed at 867 cm^{-1}; Cr^{3+}-O^{2-} stretching band is also observed at 1392 cm^{-1}; and Mg-O stretching band is also observed at 512 cm^{-1} [44–51].

The FTIR spectrum of sintered HCFC slag sintered at 1200 °C is shown in Figure 5.10b. In Figure 5.10b, broad stretching bands can be seen around 3451 cm^{-1} and 2941 cm^{-1}. These bands arise due to the stretching vibrations of O-H bonds and H bending vibrations of H-O- H of interlayer adsorbed H_2O molecule. C=O stretching band is indicated at 2248 cm^{-1} and confirms the presence of carbonate groups. The main reason for this being the presence of chemisorbed CO_2 in HCFC slag sintered at 1200 °C. Si-O and O-Si-O bands are also found at 1100 cm^{-1} and 618 cm^{-1}, which indicates the presence of silicate groups. Presence of Al^{3+}-O^{2-} bonds are also observed near 808 cm^{-1}. Fe-O stretching band is also observed at 468 cm^{-1}. Presence of Cr^{3+}-O^{2-} bond is also observed near 1178 cm^{-1}; and the presence of Mg-O bond is also observed near 808 cm^{-1} [44–51].

The FTIR spectrum of HCFC slag/K-feldspar (70:30) composite sintered at 1200 °C is shown in Figure 5.10c. Peak at 1113 cm^{-1} is attributed to Si-O band in SiO_4 molecules. A weak Si-O symmetrically stretching band is observed at 640 cm^{-1}. Peaks at 995 cm^{-1} and 790 cm^{-1} are assigned as alternating Si-O and Al-O bonds. Al-O stretching vibration is present in AlO_4 tetrahedra. There is shifting of stretching frequencies. It is observed that band of asymmetric bending of O-Si-O and O-Al-O band shifts to lower value and confirms the previous finding [44–51].

5.3 Conclusions

High-carbon ferrochrome (HCFC) slag can be effectively used for many engineering applications, viz., making concrete, brick production, partial replacement of cement, lightweight aggregates, ceramics, refractory materials, etc. In the present investigation, composite material made with sintering yielded high electrical resistant value. Optimum chemical composition of the ingredients was found to be 70% HCFC slag: 30% K-feldspar. With this composition, electrical resistivity was found to be highest (37.47×10^{11} Ω.cm). This value was determined after sintering at 1200 °C for 1 h. Vickers hardness value of the sintered product was estimated to be 678.95.

A SEM micrograph of sintered product displayed a compact structure. Together with microstructural analyses, EDS analyses showed thr presence of elements, such as O, Si, Al, Ca, Mg, Cr, Fe, etc., in the sintered materials. The TEM image showed fused crystal mass with some crystalline phases.

The XRD pattern identified the presence of forsterite, spinel, quartz, mullite, chromium-bearing phase, etc.

In FTIR spectra, different stretching vibrations of sintered and unsintered samples were noticed. The stretching vibration corresponds to presence of chemical groups in the test samples.

Acknowledgments

The authors would like to thank Prof. Munesh Chandra Adhikary, PG Council Chairman, Fokir Mohan University, for his invaluable guidance, advice, and constant inspiration throughout the entire process. They would also like to thank Mr. Mukteswar Mohapatra of Fokir Mohan University for his support. The authors convey their sincere thanks to GIET, University Gunupur, Rayagada, Odisha, India for providing lab facilities to do the research work. They also would like to thank CRF IIT Kharagpur for providing testing facilities.

References

1. P. Niemelä and M. Kauppi, (2007), Production, Characteristics And Use Of Ferrochromium Slags, INFACON XI , Vol. xxx, pp. -one hundred seven- one hundred seventy nine.
2. J. Zelic, (2005), Properties of Concrete Pavements Prepared with Ferrochromium Slag as Concrete Aggregate, Cement and Concrete Research, Vol. 35, pp. 2340-2349 DOI: 10.1016/j.cemconres.2004.11.019
3. C. Panda, K. K. Mishra, Kishor Chandra Panda, B.D. Nayak and B.B. Nayak, (2013), Environmental and Technical Assessment of Ferrochrome Slag as Concrete Aggregate Material, Construction and Building Materials, Vol. 49, pp. 262-271 DOI: 10.1016/j.conbuildmat.2013.08.002)
4. O. Gencel, M. Sutcu, E. Erdogmus, V. Koc, V. V. Cay, and M. S. Gok, (2013), Properties of Bricks with Waste Ferrochromium Slag and Zeolite, Journal of Cleaner Production Vol. 59, pp. 111-119 (http://dx.doi.org/10.1016/j.jclepro.2013.06.055).
5. S. Zhang, L. Liu. K. Tan, L. Zhang, and K. Tang, (2015), Influence of Burning Temperature and Cooling Methods on Strength of High Carbon Ferrochrome Slag Lightweight Aggregate, Construction and Building Materials, Vol. 93, pp. 1180-1187 (https://doi.org/10.1016/j.conbuildmat.2015.04.045).
6. O. Gencel, M. Sutcu, E. Erdogmus, V. Koc, V. V. Cay, and M. S. Gok, (2013), Properties of Bricks with Waste Ferrochromium Slag and Zeolite, Journal of

Cleaner Production, Volume 59, 15 November, Pages 111-119 (https://doi.org/10.1016/j.jclepro.2013.06.055).

7. H. Li, C. Li, L. Wu, H. Wang, J. Li, M. Fu, C.-A. Wang, (2020), Near net size sintering of porous cordierite ceramics with excellent properties, Journal of Alloys and Compounds, Vol. 826, pp. 154121-xxx (https://doi.org/10.1016/j.jallcom.2020.154121).
8. W. Kronert and U. Schumacher, (1989), Use of Low-Cement and Ultralow-Cement Refractory Castables in the Iron and Steel Industry Furnace Prospects, Interceram (Aachen Proc. 1988) Vol. 3, pp. 12–18.
9. H. Peng, B. Myhre and J.-O. Riszinski, (2007), Proc. 5th international Symposium on Refractories (ISR 2007), Beijing, China, April 11–12.
10. B. Myhre, (1994), Strength Development of Bauxite-Based Ultralow-Cement Castables, American Ceramic Society Bulletin, Vol. 73, pp. 68–73.
11. B. Myhre, (1993), 95th Annual Meeting and Exposition of the American Ceramic Society, Cincinnati, OH 18–22 April 1993,
12. K. Xianxin Zhou, Sankaranarayanane and M. Rigaud, (2004), Design of Bauxite-based Low-Cement Pumpable Castables: A Rheological Approach, Ceramic International, Vol. 30, pp. 47–55.
13. H. Motz and J. Geiseler, (2001), Products of Steel Slags an Opportunity to Save Natural Resources, Waste Management, Vol. 21, pp. 285–293.
14. B.B. Lind, A.M. FaÈllman and L.B. Larsson, (2001), Environmental Impact Of Ferrochrome Slag In Road Construction, Waste Management, Vol. 21, pp. 255–264.
15. D.S. Rao, S.I. Angadi, S.D. Muduli and B.D. Nayak, (2010), Recovery of Chromite Values from Ferrochrome Industry Flue Dust" AT Mineral Processing, At Mineral Processing Engl Ed, Vol. 51, pp. 1–5
16. A. Yılmaz and M. Karasahin, (2010), Mechanical Properties of Ferrochromium Slag in Granular Layers of Flexible Pavements, Materials and Structures, Vol. 43, pp. 309–317.
17. J. Zelic, (2005), Properties of Concrete Pavements Prepared with Ferrochromium Slag as Concrete Aggregate, Cement and Concrete Research, Vol. 35, pp. 2340–2349.
18. A. Yılmazand and M. Karasahin, (2011), Journal of ASTM International, Vol. 9.
19. Ideas 1st Research is registered trademark of ideas 1st information service limited (www.ifisresearch.com).
15. X. Pan, (2013), Mineral Processing and Metallurgical Engineering (ICMMME'2013), Johannesburg, South Africa, pp.15–16,
16. L. Holappa and Y. Xiao, (2004), Slags in ferroalloys production - Review of present knowledge, Journal-South African Institute of Mining and Metallurgy, Vol. 104, pp. 429–438.
17. H. Mehmet Erdem, M. Soner Altundogan, D. Turan and F. Tumen, (2005). Hexavalent Chromium Removal by Ferrochromium Slag, Journal of Hazardous Materials, Vol. 126, pp. 176–182.

18. G.M. Biggar and M.J. O'Hara, (1970), Retrograde Solubility of Periclase, Forsterite, and Dicalcium SilicateJournal of the American Ceramic Society, Vol. 53, pp. 538-540.
19. https://www.metals-hub.com/blog/ferrochrome-plants-worldwide-and-their-annual-production-capacity/
20. D.-G. Zhu, C. Yang, J. Pan, L. Lu, Z. Guo, and X. Liu, (2018), An Integrated Approach for Production of Stainless Steel Master Alloy From a Low Grade Chromite Concentrate, Powder Technology, Vol. 335, pp. xxx-xxx (DOI: 10.1016/j.powtec.2018.05.018).
21. I. N. Murthy and J. B. Rao, (2016) Investigations on Physical and Chemical Properties of High Silica Sand, Fe-Cr Slag and Blast Furnace Slag for Foundry Applications, Procedia Environmental Sciences, Vol. 35, pp. 583–596 (doi: 10.1016/j.proenv.2016.07.045).
22. J.-L. Li, A.-J. Xu, D.-F. He, Q.-X. Yang, and N.-Y. Tian, (2013), Effect of FeO on the Formation of Spinel Phases and Chromium Distribution in the CaO-SiO_2-MgO-Al_2O_3-Cr_2O_3 System International Journal of Minerals, Metallurgy and Materials Vol. 20, pp. 253-xxx (DOI: 10.1007/s12613-013-0720-9).
23. Z. A, Babakhanova and M. Kh Aripova (2021), Slag Resistant and Refractory Ceramic Compositions in MgO – Al_2O_3 – SiO_2 System on the base of Local Raw Materials of Uzbekistan, IOP Conf. Series: Materials Science and Engineering, Vol. 1091 pp. 012057-xxx (doi:10.1088/1757-899X/1091/1/012057).
24. H. Zuo, R. Liu, and A. Lu, (2022), The Behavior of Water in Orthoclase Crystal and Its Implications for Feldspar Alteration, Crystals, Vol. 12, pp. 1042-12 (https://doi.org/10.3390/cryst12081042).
25. E. Hinze, (1982) Laboratory Electrical Conductivity Measurements on Mantle Relevant Minerals, Geophysical Surveys, Vol. 4, pp. 337-352 (0046-5763/82/0044-0337502.40).
26. H. Hu, H. Li, L. Dai, S. Ming Shan, and C. Zhu, (2011), Electrical Conductivity of Albite at High Temperatures and High Pressures, American Mineralogist, Vol. 96, pp. 1821-1827 (DOI: 10.2138/am.2011.3796).
27. H. Ni, H. Keppler, G. Manthilake, and T. Katsura, (2011), Electrical Conductivity of Dry and Hydrous $NaAlSi_3O_8$ Glasses and Liquids at High Pressures, Contributions to Mineralogy and Petrology, Vol.162, pp. 1-13 (DOI: 10.1007/s00410-011-0608-5).
28. X. Yang, H. Keppler, C. Mccammon, and H. Ni, (2012), Electrical Conductivity of Orthopyroxene and Plagioclase in the Lower Crust, Contributions to Mineralogy and Petrology, Vol. 163, pp. 33-48 (DOI: 10.1007/s00410-011-0657-9).
29. H. Hu, H. Li, L. Dai, S. Shan, and C. Zhu, (2013), Electrical Conductivity of Alkali Feldspar Solid Solutions at High Temperatures and High Pressures, Physics and Chemistry of Minerals, Vol. 40, pp. 51–62 (DOI 10.1007/s00269-012-0546-4).
30. J. C. Egbai, (2013), Kaolin Quantification in Ukwu-nzu and Ubulu-uku using Electrical Resistivity Method, International Journal of Recent Research and Applied Studies, Vol. 14, pp. xxx-xxx.

31. O. A. Babalola and T. Akomolafe, (2008), Effects of Kaolin Particle Size and Annealing Temperature on the Resistivity of Zinc-Kaolin Composite Resistors, Journal Applied Science Environmental Management, Vol. 12, p. 113-xxx.
32. L. Li and Y. Li, (2017), Development and Trend of Ceramic Cutting Tools from the Perspective of Mechanical Processing, IOP Conf. Series: Earth and Environmental Science, Vol. 94, pp.1-6.
33. M. Zhu, R. Ji, Z. Li, H. Wang, L. L. Liu, and Z. Zhang, (2016), Preparation of Glass Ceramic Foams for Thermal Insulation Applications from Coal Fly Ash and Waste Glass, Construction and Building Materials, Vol. 112, , pp. 398-405.
34. W. V. Siemens, (1966), Inventor and Entrepreneur: Recollections of Werner Von Siemens, London, England,.
35. T. Tanaka and T. Imai, (2013), Advances in Nanodielectric Materials Over the Past 50 years, IEEE Electrical Insulation Magazine, Vol. 1, pp. 10-23.
36. T. Tanaka, (2005), Dielectric Nanocomposites with Insulating Properties, IEEE Transactions on Dielectrics and Electrical Insulation, Vol. 12, pp. 914-928.
37. https://www.prnewswire.com/news-releases/ceramic-capacitors-technologies-and-opportunities-300762733.html
38. M. Panigrahi, R. I. Ganguly, R. R. Dash, (2018), An Advanced Cured High Carbon Ferrochrome Slag (HCFCS) Geopolymer (GP): A Constructional Materials in IOP Conference Series: Materials Science and Engineering, Vol. 410, pp.1-15.
39. H. Hu, H. Li, L. Dai, S. Shan, and C. Zhu, (2013), Electrical Conductivity of Alkali Feldspar Solid Solutions at High Temperatures and High Pressures, Physics and Chemistry of Minerals, Vol. 40, pp. 51-62 (DOI 10.1007/s00269-012-0546-4).
40. M. K. Panigrahi, (2021), Investigation of Structural, Morphological, Resistivity of Novel Electrical Insulator: Industrial Wastes, Bulletin of Scientific Research, Vol. 3, pp. 51-58.
41. M. K. Panigrahi, (2022), Investigation of Structures of Sintered Fly Ash Materials: Resources of Industrial Wastes, Bulletin of Scientific Research, Vol. 4, pp.1-10.
42. Z. Zhan, W. Sun, S. Zhang, Q. Wen, H. Zhang, and X. Xiong (2022), $(AlCrMg)_x(PO_4)_y/MgO$ composite: A New Thermal Protection and Insulation Material up to 2400 °C, Composites Part B: Engineering, Vol. 245, pp. 110198-xxx (https://doi.org/10.1016/j.compositesb.2022.110198).
43. M. K. Panigrahi, S. Mohanty, R. I. Ganguly, R. R., Synthesis And Characterizations Of High Carbon Ferrochrome (HCFC) Slag Based Geopolymer, IntechOpen Publisher, UK. ISBN: 978-1-83968-989-5), DOI: http://dx.doi.org/10.5772/intechopen.97140
44. B.J. Saikia, G.K Rao, and P. Sarathy, (2010), Fourier Transform Infrared Spectroscopic Characterization of Kaolinite from Assam and Meghalaya,

North Eastern India, International Journal of Modern Physics., Vol. 1, pp. 206.

45. R.P. dos Santos, J. Martins, C. Gadelha, B. Cavada, A.V. Albertini, F. Arruda, M. Vasconcelos, E. Teixeira, F. Alves, J.L. Filho, and V. Freire, (2014), Coal Fly Ash Ceramics: Preparation, Characterization, and Use in the Hydrolysis of Sucrose. Scientific World Journal, Vol. 2014, pp. 01.
46. A. Jose, M. R. Nivitha, M. Krishnan, and R. G. Robinson, (2020), Characterization of Cement Stabilized Pond Ash using FTIR Spectroscopy, Construction and Building Materials, Vol. 263, pp.120136-xxx (DOI: 10.1016/j.conbuildmat.2020.120136).
47. Y. Liu, F. Zeng, B. Sun, P. Jia, and I. T. Graham, (2019), Structural Characterizations of Aluminosilicates in Two Types of Fly Ash Samples from Shanxi Province, North China, Minerals, Vol. 9, pp. 358-16 (doi:10.3390/min9060358).
48. P. Rożek, M. Król, and W. Mozgawa, (2018), Spectroscopic Studies of Fly Ash-Based Geopolymers, Spectrochimica Acta - Part A Molecular and Biomolar Spectroscopy, Vol. 198, pp. 283-289 (https://doi.org/10.1016/j.saa.2018.03.034).
49. S. Kumar, F. Kristály, and G. Mucsi, (2015), Geopolymerization Behaviour of Size Fractioned Fly Ash, Advanced Powder Technology, Vol. 26, pp. 24-30 (https://doi.org/10.1016/j.apt.2014.09.001).
50. Z. Kledyński, A. Machowska, B. Pacewska & I. Wilińska, (2017), Investigation of Hydration Products of Fly Ash-Slag Pastes, Journal of Thermal Analysis and Calorimetry, Vol. 130, pp. 351-363 (https://doi.org/10.1007/s10973-017-6233-4).
51. D. Bondar and R. Vinai, (2022), Chemical and Microstructural Properties of Fly Ash and Fly Ash/Slag Activated by Waste Glass-Derived Sodium Silicate, Crystals, Vol. 12, pp. 913-12 (https:// doi.org/10.3390/cryst12070913).

6

High Resistant HCFC Slag-Based Geopolymer Materials

Muktikanta Panigrahi[1]*, Ratan Indu Ganguly[2] and Radha Raman Dash[3]

[1]Department of Materials Science, Maharaja Sriram Chandra Bhanja Deo University, Keonjhar Campus, Odisha, India

[2]Department of Metallurgical Engineering, National Institute of Technology, Rourkela, Odisha, India

[3]CSIR-National Metallurgical Laboratory (NML), Jamshedpur, Jharkhand, India

Abstract

The objective of this chapter is to explore the possibility of using high-carbon ferrochrome (HCFC) slag for developing insulators. The electrical property of geopolymer (GP) is also studied.

Pond ash (PA) is a waste material of thermal power stations with low calcium content. It has high silica (62%) and alumina (28.3%). High-carbon ferrochrome slag is a by-product of the ferrochrome industries. It contains silica (28%) and alumina (28.3%).

The mechanical properties of HCFC slag GP, such as compressive strength and hardness, are determined using several techniques. The resistance value of the geopolymer is measured using the two-probe method. Microstructures are evaluated by optical microscopy, SEM, and TEM, and different phases are determined using an X-ray diffractometer. Different chemical bonds formed during the processing of geopolymer are detected with a Fourier-transform infrared (FTIR) instrument. Thermal behavior of the material is determined by TGA. The highest compressive strength of HCFC slag GP is measured to be 19.5 MPa for an optimum treatment combination. i.e., curing temp of 70 °C and time 48 h.

Transformation of crystalline to amorphous phase is observed in the XRD pattern of GP sample. Vickers hardness value of the sample is found to be 9.12. An optical micrograph of HCFC slag GP shows the presence of particles having different sizes; and SEM images show a higher amount of glassy phase after curing for 24 h at 70 °C. Microstructure is correlated with compressive strength properties.

**Corresponding author*: muktikanta2@gmail.com

Muktikanta Panigrahi, Ratan Indu Ganguly and Radha Raman Dash (eds.) High Electrical Resistant Materials: Ferrochrome Slag Resource Ceramics, (139–158) © 2024 Scrivener Publishing LLC

A TEM micrograph of geopolymer displays fused mass structure; and the SAD pattern of HCFC slag GP indicates the presence of some crystalline phases in GP product. Phase transformation is observed during the geopolymerization process of HCFC slag. Crystalline phase is transformed to a glassy phase. The presence of desired chemical groups in the geopolymer is shown using FTIR analysis. Thermogravimetric (TG) results are also discussed for both uncured and cured GP samples.

Keywords: Industrial waste, ferrochrome slag, geopolymer, phases, compressive strength, electrical resistivity

6.1 Introduction

Solid waste generated from industry is increasing exponentially due to the increase in the number of industries. Therefore, industries are facing the problem of disposing their generated waste safely. The means and methods adopted/proposed to use the waste generated from industries will lead to sustainability. Nowadays, in the process of manufacturing cement and aggregate, many industrial by-products are used. Similarly, special concretes, such as self-compacting concrete, reactive powder concrete, ultra-high-performance concrete, etc., are developed using one or more of the industrial by-products. In line with this, special concrete called electrically conductive concrete has recently been developed using industrial by-products such as high-carbon fly ash and copper slag [1]. It is concluded in the study that the electrical conductivity of concrete can be improved by using high-carbon fly ash and copper slag. Greater than 50% replacement of copper slag improved the electrical conductivity of concrete [1].

Copper slag is a by-product obtained from the copper smelting industries [2]. The amount of slag produced during the production of copper is huge, in the range of three times that of copper manufacturing [3]. Utilization of copper slag in any useful form would reduce the piling of slag at the industrial site. Research on copper slag as fine aggregate alternative in concrete reports its outstanding mechanical and durability properties [4–6]. Furthermore, the use of copper slag reduces the drying shrinkage of concrete [7]. The improved strength and durability of copper slag incorporated concrete is reported to be due to the high toughness of copper slag [8]. Owing to the advantages reported in the literature, copper slag is considered as a sustainable and useful material for cement/concrete-based composite applications [9]. A study recently reported the influence of copper slag on improving the electrical conductivity of self-compacting concrete [10].

Electrical conductivity of cement mortar/concrete is influenced by the type and dosage of constituent materials present [11, 12]. It is reported that the presence of conductive aggregates like copper slag in mortar/concrete greatly improves the electrical conductivity [1, 13]. Furthermore, it is stated that along with the conductive aggregates, the presence of carbon content in the cement paste enhances the electrical conductivity. Conductive aggregates used for improving electrical conductivity are either artificially made, like fly ash aggregates, carbon black and carbon fiber incorporated aggregates, or by using industrial by-products like copper slag [1, 14, 15].

Portland cement (PC) is a common structural material, which is widely used in construction of structure [16–19]. Portland cement production causes environmental problems due to the release of carbon dioxide (CO_2) [20]. The production of PC is an extremely energy-intensive process [21]. Therefore, the research community has used industrial wastes, such as fly ash (FA), pond ash (PA), slag, bottom ash (BA), quarry dust (QD), etc., as resource materials for the production of geopolymer [16–20]. It is no wonder that research studies are being carried out for the replacement of PC with new-generation material, i.e., GP. Therefore, GP technology has enabled the production of an alternative material to replace PC [22, 23].

One of the wastes is slag obtained from the ferro-alloy industry during the production of ferrochrome [16–20]. The slag contains a variety of oxides, such as magnesium oxide, iron oxide and chromium oxide, which exist in more than one oxidation states. Also, ferrochrome slag may contain elemental chromium [22, 23]. The major constituents of the slag are Cr_2O_3, FeO, SiO_2, MgO, Al_2O_3, and CaO [22, 23]. In the year 1978, Davidovits described geopolymer to be a future construction material which was predicted to replace PC in construction areas [24]. Reports are available which indicate waste materials, i.e., fly ash, bottom ash, pond ash, slag, etc., are suitable for production of GP [25–28]. These oxides form chain-like structure by the reaction with active alkali solution such as sodium hydroxide (NaOH) or potassium hydroxide (KOH) in combination with sodium silicate [27, 29]. Thus, GP can be considered as an innovative construction material for the 21st century [24–29]. Many researchers have described geopolymer structure [30, 31]. The advantages of using GP materials lie in the fact that they are cheaper, non-hazardous, fire resistant and environmentally friendly [24–29].

The corrosion test and pH measurements have revealed that GP material has low leachability and excellent frictional resistance. Therefore, it can be used for construction of roads [32–35].

In the present investigation, HCFC slag is selected as a raw material to prepare geopolymer. Effort has been made to characterize (such as by mechanical, structural, morphological, thermal, and electrical analyses) raw HCFC slag and cured HCFC slag geopolymer in different lengths of curing time.

6.2 Experiment Details

6.2.1 Chemicals and Materials

In the preparation of geopolymer, HCFC slag was chosen as a raw material and was procured from a ferrochrome industry (Shyam Ferro Alloys, Durgapur, West Bengal, India). Table 6.1 lists the compositions of raw HCFC slag. Sodium hydroxide (NaOH) and sodium silicate (Na_2SiO_3) were purchased from Loba Chemie, India, and Merck, India, respectively. NaOH acts as an activator and Na_2SiO_3 acts as a polymerization agent. Sika chemical was procured from a reputed firm at the Visakhapatnam market in Andhra Pradesh, India. It was used to enhance plasticity of geopolymer.

6.2.2 Synthesis of Geopolymer from HCFC Slag

Consecutive steps for synthesizing geopolymer from HCFC slag are described below [27, 28]. A flow chart of the entire geopolymerization process is shown in Figure 6.1.

Step 1: Raw material preparation
HCFC slag is received in the form of lumps. The lumps are crushed and ground in a ball mill. Thereafter, it is sieved with 240 meshes. Fine particles are dried in an oven (2 h at 120 °C). Such slag is known as processed slag.

Alkaline activator is a mixture of sodium silicate and sodium hydroxide in optimized ratio. The solution is continuously stirred in a magnetic

Table 6.1 Chemical composition of HCFC slag by percent (%) [28].

Constituents of HCFC slag with percentages	SiO_2 = 28–31 MgO = 26.0–26.5 Al_2O_3 = 23.5–24.5 CaO = 7.5–8.5 Cr_2O_3 = 4.5–5.4 FeO = 2.5–3.0

Source: Shyam Ferro Alloys, Durgapur, India.

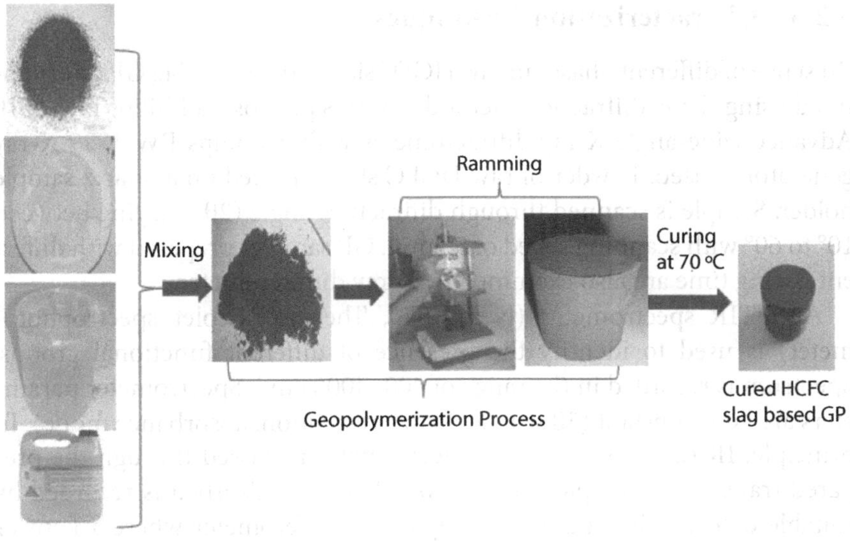

Figure 6.1 Flow chart showing the entire geopolymerization process [28].

stirrer. Plasticizer (Sika) with extra water (a few percentages) is poured into the liquid mixtures.

Step 2: Geopolymer preparation with raw materials
Processed HCFC slag is poured into a mixture container. Then, alkali activator solution is added slowly to the slag and is manually mixed. Such mixture is left for five minutes to complete polymerization reaction.

Step 3: Molding, casting and compaction
In this step, a cylindrical shaped mold is used having 70 mm height and 55 mm diameter. The poured pasty material is rammed continuously 15 times. It is then left for 3 minutes under compaction before demolding samples. Sample is collected, and 10 similar samples are prepared in order to get average strength value. Collected sample is called green sample. It should be mentioned that their inside surfaces (i.e., cylindrical shaped molds) are coated with diesel oil to ensure smoother surfaces of test sample.

Step 4: Curing
Green samples are kept for heating in an oven for 24 h at 70 °C followed by curing in open air (7 days and 28 days) [36, 37].

Cured specimen is used for different investigations, i.e., XRD, FTIR SEM, TEM, TGA, DSC, compressive strength, and resistivity.

6.2.3 Characterization Techniques

Presence of different phases in raw HCFC slag and HCFC slag GPs are identified using X-ray diffraction method. For this purpose, a Philips PW-1710 Advance wide-angle X-ray diffractometer with a Philips PW-1729 X-ray generator is used. Powder of raw HCFC slag is placed on a quartz sample holder. Sample is scanned through diffraction angle (2θ) varying between 10° to 60° with scanning speed of 5°/min. GP samples prepared with different curing time are also examined by X-ray diffractometer.

An FTIR spectrometer (Nexus-870, Thermo Nicolet spectrophotometer) is used to identify the presence of different functional groups. Spectrum is recorded in the range of 400–4000 cm^{-1}. Spectrometer parameters are kept constant (50 scan at 4 cm^{-1} resolution, absorbance mode). In principle, IR radiation from the spectrometer is passed through the prepared transparent sample. Amount of IR energy adsorbed is recorded by suitable detector. It is guided through an interferometer where a Fourier transform is performed on the output signal. Transparent FTIR pellet (13 mm diameter, 0.3 mm thick) is prepared with small amount of test samples and bulk amount of potassium bromide (KBr). A background spectrum is collected before running the samples.

Surface morphologies of raw HCFC slag and HCFC slag GP are analyzed by scanning electron microscope (SEM). A Carl Zeiss Supra-40 microscope is employed to analyze surface morphologies. A scanning electron microscope (SEM) uses secondary electrons in place of light to produce an image. In the microscope, test specimen is kept in a vacuum. Generated electron beam is focused on a small portion of the sample. The detector collects the output signals generated by interaction of and is displayed on a screen. During the sample preparation, a special method is used to avoid moisture absorption. Since test samples are non-conducting in nature, they are gold-coated by sputtering technique. Operating voltage of gold coating (by sputtering technique) is kept at 4 kV. Elemental analyses are done with EDX, which is attached to the scanning electron microscope.

A TEM is special kind of electron microscope. It uses highly accelerated beam of electron (as illumination) during their analyses. Such beams are passed through the electron transparent samples. Interaction occurs between electron and atoms of the sample. The transmitted electrons are then focused by the objective lens into an image. In TEM, electromagnetic lenses are used to guide the electron beam through the samples. The specimen preparation for TEM experiment is an important step for getting good results. The main criterion for specimen preparation is artifact-free sample. The detailed sampling procedure is described in the literature [16].

In this investigation, pellet samples are used for TEM. An SAD pattern can be obtained along with microscopic studies in TEM of raw HCFC slag and HCFC slag geopolymer. These studies are undertaken with a TEM instrument (5022/22, Tecnai G220 S-Twin, Czech Republic) used to carry out the measurement. TEM specimen (HCFC slag) is prepared by drop casting technique. TEM specimen of HCFC slag GP is prepared by micro-tone technique using Leica Microsystems GmbH A-1170. Then, specimens are transferred to carbon-coated Cu TEM grids.

Thermogravimetric analysis records change in mass with temperature of test samples. During heating, volatile materials present in test materials are eliminated due to input thermal energy. Therefore, degradation behavior of materials or thermal stability of materials are understood by conducting this experiment. In the present investigation, HCFC slag and HCFC slag GP are used. A Perkin Elmer Pyris Diamond analyzer is used for this investigation. TGA analyses are carried out in N_2 gas environment and TGA run is taken in the range of 50–300 °C.

Compression test of HCFC slag GP is performed using an Ultimate Tensile machine (AIMIL COMPTEST 2000 model). Standard ASTM is followed to determine compressive strength value. Sample used is cylindrical shaped with 100 mm diameter and 200 mm length. Stress-strain curves are plotted by keeping strain rate value low. Maximum load is taken at the point of fracture of the sample.

Electrical resistivity value of cured GP pellet is determined by the following equation [38, 39]

$$\rho = \left(\frac{R \times A}{l} \right) \tag{6.1}$$

Here, A is the area of cross section of sample $= \frac{\pi}{4} D^2$ (cm^2), R is the resistance of prepared samples (Ω), l is thickness of the sample, ρ is the resistivity of the samples in Ω.cm.

6.2.4 Results and Discussion

Ferrochrome slag is obtained as waste material during production of ferrochrome alloy in a smelter. Pouring temperature of slag is around 1500 °C. The slag contains oxides of magnesium, iron and chromium in their different oxidation states. Chromium present in the slag may either be trivalent (+3 oxidation state) and or hexavalent (+6 oxidation state). Some elemental chromium may also be present in the slag. The oxides are usually spinel phase (i.e., $FeCr_2O_4$). $MgAl_2O_4$ spinel phase is also in the slag. The presence of these spinel phases helps the slag to enhance mechanical, chemical and thermal properties both at ambient and elevated temperatures [40, 41]. Table 6.1 shows chemical analysis of ferrochrome slag. It is evident that SiO_2, Al_2O_3, Fe_2O_3, and Cr_2O_3 are present in major quantities, while CaO, MgO, and FeO are present in minor quantities [42]. The ground and sieved (240 meshes) of as-received slag is dried for 2 h at 120 °C for removal of moisture. The sieved HCFC slag is ready for geopolymer preparation.

Geopolymer is produced by reaction of HCFC slag with concentration of sodium hydroxide (8 M), sodium silicate, water-soluble plasticizer (i.e., Sika). Compressive strength of cured samples are indicated in Table 6.2. Samples are cured for different lengths of curing time (i.e., 4 h, 16 h, 18 h, 24 h) at 70 °C. This is done to maximize strength of the samples for an optimized curing temperature. Table 6.2 clearly shows rapid increase in compressive strength of GP with increasing curing time up to 18 h. Between 24 h and 48 h curing, the increase of strength is marginal. Therefore, curing time is kept at 24 h to minimize curing time and processing cost [43–46].

Resistance values of HCFC slag and HCFC slag geopolymers are determined by two-probe method. Estimated resistivity values of raw HCFC slag and HCFC slag geopolymers are shown in Table 6.3. It is observed that the highest resistivity value is found to be 12.72×10^9 Ohm.cm for HCFC slag geopolymer cured at 70 °C and 24 h. The lowest resistivity value is estimated to be 0.007×10^9 Ohm.cm for HCFC slag geopolymer cured at 70 °C and 4 h. The difference in resistivity value is due to the formation of different extent of geopolymer [47–50]. Curing time has played an important

Table 6.2 Comprehensive strength value of raw HCFC slag and HCFC slag geopolymer prepared with optimum treatment combination with different curing times.

S.No.	Sample ID	Compressive strength value (MPa)
1	S1 (Pure HCFC Slag powder)	---
2	S2 (HCFC Slag GP cured at 4 h)	3.0
3	S3 (HCFC Slag GP cured at 16 h)	11.0
4	S4 (HCFC Slag GP cured at 18 h)	14.0
5	S5 (HCFC Slag GP cured at 24 h)	19.0
6	S5 (HCFC Slag GP cured at 48 h)	19.7

Table 6.3 Resistivity value of raw HCFC slag and HCFC slag geopolymer prepared with optimum treatment combination with different curing time.

S.No.	Sample ID	Resistance	Resistivity (ρ) in Ohm.cm
1	S1 (Pure HCFC Slag powder)	---	---
2	S2 (HCFC Slag GP cured at 4 h 70 °C)	0.5 MΩ	0.007×10^9
3	S3 (HCFC Slag GP cured at 16 h 70 °C)	16.3 MΩ	0.17×10^9
4	S4 (HCFC Slag GP cured at 18 h 70 °C)	19.8 MΩ	0.28×10^9
5	S5 (HCFC Slag GP cured at 24 h 70 °C)	0.9 GΩ	12.72×10^9

role since it has enhanced resistivity of HCFC slag GP cured at 70 °C. Time of curing where extent of polymerization has influenced kinetics of polymerization.

Figure 6.2(a–d) shows superimposed XRD pattern of raw HCFC slag and HCFC slag-based GP materials cured at 70 °C for different curing times (4 h, 18 h, and 24 h). Sharp prominent XRD peaks of raw HCFC slag are observed (Figure 6.2a). Identified phases are forsterite, spinel (magnesium aluminate), and clinopyroxene, respectively (JCPD file). From Figure 6.2(b–d), it is observed that there are some new peaks and new peak heights are found. This has happened due to geopolymerization reaction. Also, from the pattern it is found that raw HCFC slag has crystalline phases, whereas the highest resistivity value of geopolymer has both crystalline and non-crystalline glassy phases. Peak broadening in the XRD pattern of geopolymer showed the highest resistivity value. This is formed due to chemical and thermal treatment during polymerization reaction. Line broadening and decreasing peak height indicates the occurrence of chemical reaction during polymerization.

Table 6.4 shows the mechanical properties of HCFC slag geopolymer (GP). Measurement is recorded by an instrument for different strength properties. For obtaining strength values, nanoindentation technique is used. Vickers hardness value of the test specimen is found to be 9.12. Young's modulus (EIT) and reduced modulus (E^*) are found to be 15.35 MPa and 16.868 GPa, respectively.

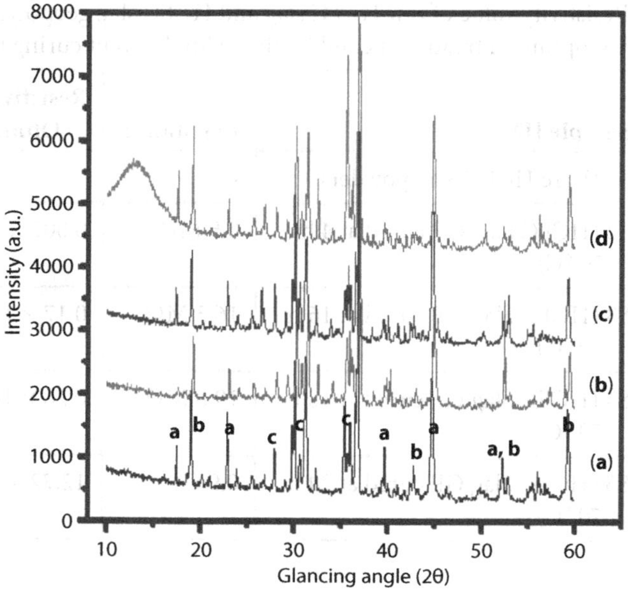

Figure 6.2 The XRD pattern of pure HCFC slag (a); HCFC slag-based GP 4 h 70 °C (b); HCFC slag-based GP 18 h 70 °C (c); HCFC slag-based GP 24 h 70 °C (d) [a=forsterite, b=spinel (magnesium aluminate), c=clinopyroxene].

Table 6.4 Vickers hardness value of HCFC slag geopolymer.

S.No.	Sample ID	F_{max} (mN)	HIT (MPa)	HVIT	EIT (MPa)	E* (GPa)
1	T1	98.97	98.452	9.1178	15.35	16.868

Note: T1 = HCFC slag of geopolymer prepared with optimum treatment combination and cured at 70 °C for 24 h.

Figure 6.3 shows an optical image of HCFC slag geopolymer obtained from samples cured at 24 h and 70 °C. Figure 6.3 displays a mixture of irregular shaped white patches, black-brown spots, and black spots spread throughout the background. Because of poor contrast, it is not possible to describe the actual features present in the microstructure. Therefore, electron spectroscopic studies (i.e., SEM or TEM) are carried out. This will enable a better analyses of microscopic features.

The SEM images of raw HCFC slag and HCFC slag GP samples cured for different curing lengths of time are studied (Figure 6.4a–d). The SEM micrograph of HCFC slag in Figure 6.4a shows crystalline phase obtained

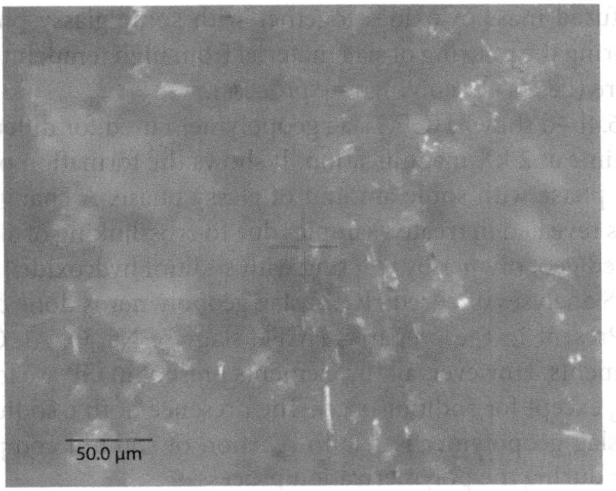

Figure 6.3 Optical image of HCFC slag geopolymer obtained from optimum treatment combination (cured for 24 h at 70 °C).

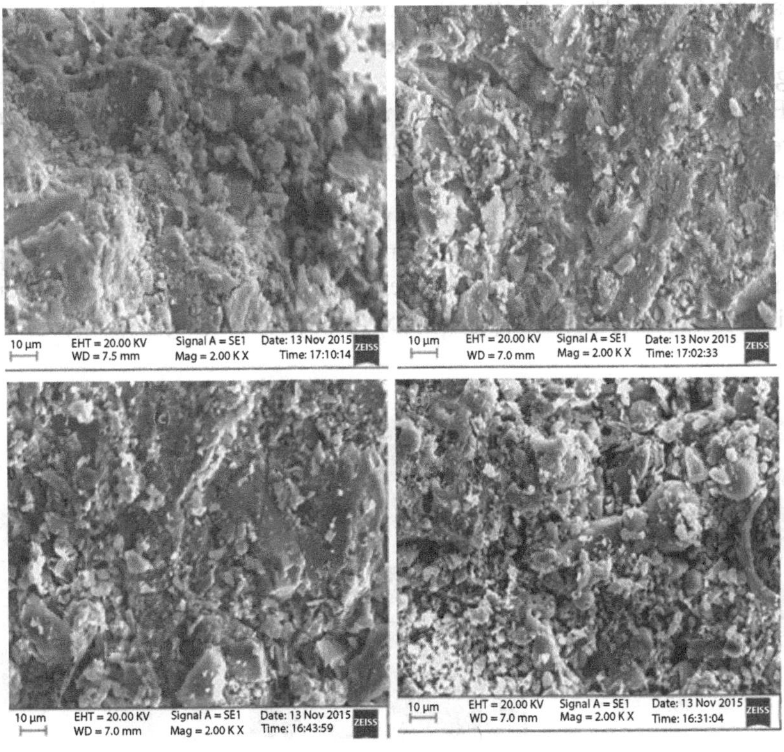

Figure 6.4 HCFC pure slag, 2 kX (a); HCFC slag-based GP 4 h 70 °C, 2k X (b), HCFC slag-based GP 18 h 70 °C, 2 kX (c); HCFC slag-based GP 24 h 70 °C, 2kX (d).

from the fused mass of oxides together with some glassy phase. This is formed during the pouring of slag material from high temperature to room temperature (i.e., faster cooling rate process).

Figure 6.4b–d shows HCFC slag geopolymer cured for different lengths of curing time at 2 kX magnification. It shows the formation of more new crystalline phase with some amount of glassy phase. A change in microstructure is revealed in treated samples due to crosslinking of alumino-silicate in cured geopolymer by reacting with sodium hydroxide [51].

The EDS analyses of cured HCFC slag geopolymer is done during SEM analyses. Present in the prepared HCFC slag are Na, O, Al, Ca, Mg, Cr, and Fe elements. However, all the elements present in GP are found in raw HCFC slag except for sodium [xxx]. The presence of th e sodium element in HCFC slag geopolymer is due to reaction of sodium compound with HCFC slag during the polymerization process.

The HRTEM image (a), SAD pattern (b), and lattice pattern (c) of HCFC slag geopolymer cured at 70 °C for 24 h are shown in Figure 6.5. Figure 6.5a shows fused mass product of small irregular shaped crystal. This mass product is obtained after the completion of polymerization process. Similar mass product is also observed in the SEM micrograph of HCFC slag geopolymer (Figure 6.4b–d).

Figure 6.5b shows the diffraction pattern of HCFC slag geopolymer cured at 70 °C for 24 h. It shows bright spots in different sizes. Spots are arranged in different orientation in different circumferences around the center. Spots indicate crystallinity of phases present in materials.

Lattice pattern of the TEM specimen is shown in Figure 6.5c. The pattern shows lattice planes having different orientation. It signifies the crystalline nature and is well-supported by SAD pattern (Figure 6.5b).

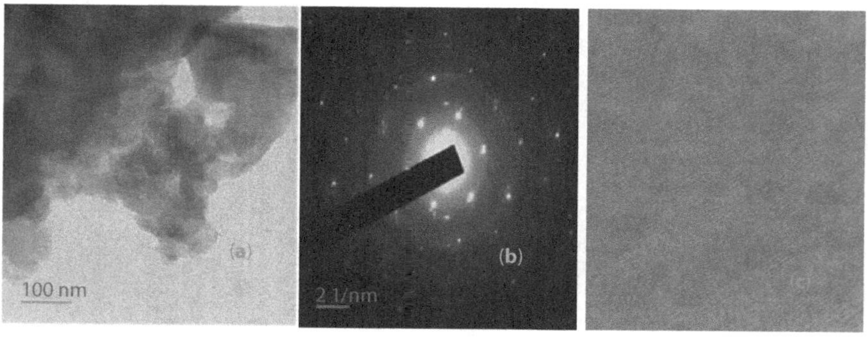

Figure 6.5 HRTEM analysis of HCFC slag geopolymer cured at 70 °C and 24 h [a=TEM image, b=SAD pattern, and c=lattice pattern].

Superimposed FTIR spectra of HCFC slag and HCFC slag geopolymer cured for different lengths of time (at 70 °C) are shown in Figure 6.6. The entire FTIR spectra indicate presence of different absorption bands of indicated material (Figure 6.6). Figure 6.6a shows absorption bands occurring at different wave numbers, i.e., 3438, 2928, 1648, 1411, 1079, 872, and 512 cm^{-1}. Bands shown in Figure 6.6a correspond to O-H bonds (3438 cm^{-1} wave number), H-O-H bending vibrations (1648 cm^{-1} wave number) of interlayer adsorbed H_2O molecule [52–56], Si-O band (at 1079 cm^{-1} wave number), $Al^{3+}O^{2-}$ absorption (at 872 cm^{-1} wave number) [52–56], Fe-O band (at 512 cm^{-1} wave number) [52–56], respectively. The FTIR spectrum of as-received HCFC slag is compared to those of other reacted products. Striking difference is observed. Peaks of bands have shifted to higher wave number with increase in absorption values. Around 1075 cm^{-1}wave number, higher absorption value is found. There is a sharp drop in absorption values for all reacted products neighboring wave number 1750 cm^{-1}. There is a sharp rise in absorption value for increase in wave number. Around 1000 cm^{-1}, peak height has reached maximum. Peaks are observed for all the reacted products. Thus, alkali treatment of HCFC slag has caused structural changes in materials which is due to geopolymerization of raw materials by alkali treatment and curing.

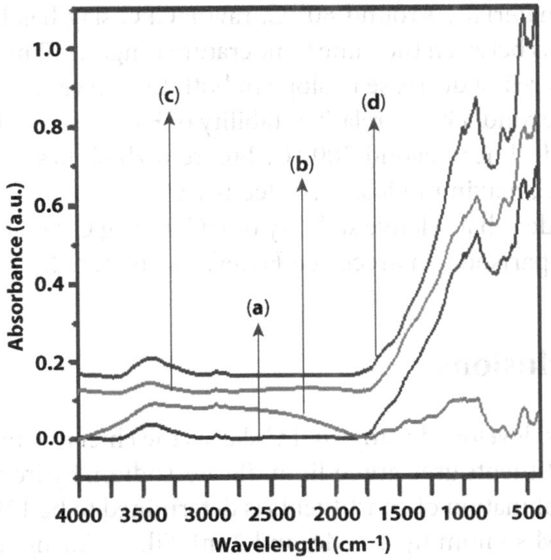

Figure 6.6 FTIR spectra of raw HCFC slag (a); HCFC slag-based GP 4 h 70 °C (b); HCFC slag-based GP 18 h 70 °C (c); HCFC slag-based GP 24 h 70 °C (d).

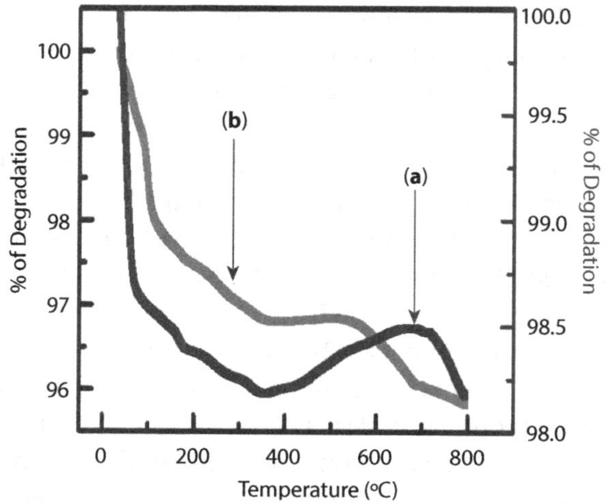

Figure 6.7 TGA plot of HCFC slag (a), and HCFC slag GP cured for 24 h at 70 °C (b) [57].

Figure 6.7 shows superimposed TGA curves of HCFC slag and HCFC slag GP (curing time ~24 h and curing temperature ~70 °C). In this plot, Y-axis on the left represents percentage of weight loss (%) for as-received HCFC slag materials, whereas Y-axis on the right shows percentage of weight loss (%) for as-received HCFC slag GP material. It can be seen that there is a sharp drop in weight loss (%) of HCFC slag materials in comparisons to GP materials. Around 80 °C, raw HCFC slag has lost weight by ~2.8%, whereas between the same temperature ranges, GP material has lost ~1%. [57]. There is a decrease in slope in both the curves indicating rate of weight loss. Beyond 350 °C, relative stability of HCFC slag is increased and then rapidly declines beyond 700 °C. Figure 6.7b shows a similar trend. Beyond 550 °C, maximum loss recorded is 1.9%.

It is concluded that relative stability of HCFC slag GP material is always higher in comparison to as-received HCFC slag material.

6.3 Conclusions

Geopolymer is developed using HCFC slag as the chief raw material. Other chemical constituents used are sodium silicate, sodium hydroxide, and Sika. Optimum combination of constituents is determined to be 12% sodium silicate, 3% of 8M sodium hydroxide, and 2 mL Sika. Curing temperature is fixed. Curing time is varied from 4 h, 16 h, 20 h and 24 h. It is observed that strength value of the material increased with increasing curing time.

The maximum strength value of GP is found to be 19 MPa for optimum combination of constituents. Vickers hardness value is measured to be 9.12.

An optical micrograph showed the presence of different shaped particles. Microstructure (i.e., SEM image) showed crystalline and non-crystalline phases. Distinct changes were observed for GP material if compared with HCFC slag. Changes of microstructure have resulted due to reaction occurring during geopolymerization of HCFC slag. The TEM image of geopolymer indicated the presence of fused mass product. The SAD pattern revealed the existence of crystalline phase in the matrix of fused mass.

Phase transformation was also observed in XRD pattern. Crystalline phase of HCFC slag transformed to glassy phase during geopolymerization.

Superimposed FTIR spectra indicated a greater number of peaks in GP material if compared with HCFC sample. This explains the formation of new bonds in geopolymer material.

The TGA plots revealed degradation behavior occurring in GP and slag materials within the temperature range of 50–800 °C.

The highest resistivity value was determined to be 12.72×10^9 (Ohm.cm) by two-probe method.

Acknowledgments

The authors convey their sincere thanks to the Ministry of Mines, Government of India, for providing the grant to carry out the work (Grant number = F.No.:14/54/20214-Met.-IV; dated: 29.12.2014). The authors also convey their sincere thanks to GIET, University Gunupur, Rayagada, Odisha, India for providing lab facilities to do the research work. They would also like to thank CRF IIT Kharagpur for providing testing facilities. The authors would like to thank Prof. Munesh Chandra Adhikary, PG Council Chairman, Fokir Mohan University, for his invaluable guidance, advice, and constant inspiration throughout the entire process. They also would like to thank Mr. Mukteswar Mohapatra of Fokir Mohan University for his support.

References

1. S. Lim, W. Lee, H, Choo, and C. Lee, (2017), Utilization of High Carbon Fly Ash and Copper Slag in Electrically Conductive Controlled Low Strength Material, Construction and Building Materials, Vol. 157, pp. 42–50.

2. M. Najimi, J. Sobhani and A. R. Pourkhorshidi, (2011), Durability of Copper Slag Contained Concrete Exposed to Sulfate Attack, Construction and Building Materials, Vol. 25, pp. 1895–1905.
3. B. Ali, G. M. Modiri, A. F. Gozali, and A. Mahmoud, (2015), Effects of Copper Slag and Recycled Concrete Aggregate on the Properties of CIR Mixes with Bitumen Emulsion, Rice Husk Ash, Portland Cement and Fly Ash, Construction and Building Materials, Vol. 96, pp.172–180.
4. X. Wang, D. Geysen, S.V. Padilla, D. Tinoco, N. Hoker, T. Van Gerven, and B. Blanpain, (2015), Characterisation of Copper Slag in View of Metal Recovery, Mineral Processing and Extractive Metallurgy, Vol. 124, pp. 83–87.
5. S. Goni, P. Ma Lorenzo, and J. L. Sagrera, (1994), Durability of Hydrated Portland Cement with Copper Slag Addition in NaCl Na_2SO_4 medium, Cement and concrete research, Vol. 24, pp. 1403–1412.
6. Lye Chao-Qun, Koh Siew-Kiang, Mangabhai Raman and Dhir Ravindra K (2015), Use of Copper Slag and Washed Copper Slag as Sand in Concrete: A State-of-the-Art Review, Magazine of Concrete Research, Vol. 67, pp. 665–679.
7. N. Gupta, (2020), Strength and Durability Properties of Self-Compacting Concrete Incorporating Copper Slag. PhD thesis, Department of Civil Engineering, Thapar University, Punjab, India.
8. S. Al-Jabri Khalifa, H. Al-Saidy Abdullah, and T. Ramzi, (2011), Effect of Copper Slag as a Fine Aggregate on the Properties of Cement Mortars and Concrete, Construction and Building Materials, Vol. 25, pp. 933–938.
9. A. Krishna, P. Saeed, and Z.. Lianyang, (2015), Alkali Activation of Copper Mine Tailings and Lowcalcium Flash-Furnace Copper Smelter Slag, Journal of Materials in Civil Engineering, Vol. 27, pp. 04014193-xxx.
10. R. Sharma and A. Rizwan Khan (2018), Influence of Copper Slag and Metakaolin on the Durability of Self-Compacting Concrete, Journal of Cleaner Production, Vol. 171, pp. 1171–1186.
11. W. Sihai and D. D. L. Chung, (2007), Double Percolation n the Electrical Conduction in Carbon Fiber Reinforced Cement-based Materials, Carbon, Vol. 45, pp. 263–267.
12. W. Jianmin, L. Jianguo, and Y. Fei, (2015), Three-Phase Composite Conductive Concrete for Pavement Deicing, Construction and Building Materials, Vol.75, pp. 129–135.
13. H. Yongjia, L. Linnu, J. Shun, and H. Shuguang, (2014), Conductive Aggregate Prepared using Graphite and Clay and its Use in Conductive Mortar, Construction and Building Materials, Vol. 53, pp. 131–137.
14. W. Ziyi, W. Zhi-Jie, T. Shengxuan, and H. Yunfei, (2017), Preparation and Properties of Electrically Conductive Aggregate Made using Magnetically Separated Fly Ash, Construction and Building Materials, Vol. 150, pp. 547–557.
15. C. Binmeng, L. Bo, G. Yan, L. Tung-Chai, L. Zeyu, and L. Zongjin, (2017), Investigation on Electrically Conductive Aggregates Produced by

Incorporating Carbon Fiber and Carbon Black, Construction and Building Materials, Vol. 144, pp. 106–114.
16. M. Schneider, M. Romer, M. Tschudin, and H. Bolio, (2011), Sustainable Cement Production-Present and Future, Cement and Concrete Research, Vol. 41, pp. 642-650, (http://dx.doi.org/10.1016/j.cemconres.2011.03.019).
17. D.A. Salas, A.D. Ramirez, C.R. Rodríguez, D.M. Petroche, A.J. Boero, and J. Duque-Rivera, (2016), Environmental Impacts, Life Cycle Assessment and Potential Improvement Measures for Cement Production: A Literature Review, Journal of Cleaner Production, Vol. 113, pp. 114-122 (http://dx.doi.org/10.1016/j.jclepro.2015.11.078).
18. M. Babaee and A. Castel, (2016), Chloride-Induced Corrosion of Reinforcement in Low Calcium Fly Ash-based Geopolymer Concrete, Cement and Concrete Research, Vol. 88, pp. 96-107 (http://dx.doi.org/10.1016/j.cemconres.2016.05.012).
19. B.W. Jo, J.S. Choi, K.W. Yoon, and J.H. Park, (2012), Material Characteristics of Zeolite Cement Mortar, Construction and Building Materials, Vol. 36, pp. 1059-1065 (http://dx.doi.org/10.1016/j.conbuildmat.2012.06.061).
20. C. Gunasekara, D.W. Law, S. Setunge, and J.G. Sanjayan, (2015), Zeta Potential, Gel Formation and Compressive Strength of Low Calcium Fly Ash Geopolymers, Construction and Building Materials, Vol. 95, pp. 592-599 (http://dx.doi.org/10.1016/j.conbuildmat.2015.07.175).
21. P. Duan, C. Yan, W. Zhou, and W. Luo, (2016), Fresh Properties, Mechanical Strength and Microstructure of Fly Ash Geopolymer Paste Reinforced with Sawdust, Construction and Building Materials, Vol. 111, pp. 600-610 (http://dx.doi.org/10.1016/j.conbuildmat.2016.02.091).
22. A. Petrillo, R. Cioffi, C. Ferone, F. Colangelo, and C. Borrelli, (2016), Eco-Sustainable Geopolymer Concrete Blocks Production Process, Agriculture and Agricultural Science Procedia, Vol. 8, pp. 408-418 (http://dx.doi.org/10.1016/j.aaspro.2016.02.037).
23. K.H. Yang, J.K. Song, and K. Il Song, (2013), Assessment of CO_2 Reduction of Alkali-Activated Concrete, Journal of Cleaner Production, Vol. 39, pp. 265-272 (http://dx.doi.org/10.1016/j.jclepro. 2012.08.001).
24. An article published by the Commission of the European Communities in 1982, outlines the reasons why the generic term geopolymer was chosen for this new chemistry. See: J. Davidovits, The Need to Create a New Technical Language For the Transfer of Basic Scientific Information, in Transfer and Exploitation of Scientific and Technical Information, Proceedings of the symposium, Luxemburg, 10–12 June 1981, pp. 316-320. It is available as a pdf-file and may be downloaded from the European Parliament Bookshop. Go to < https://publications.europa.eu/en/publication-detail/-/publication/02a1db8b-3873-46d7-9e72-a6e02660e154 >
25. X. Y. Zhuang, L. Chen, S. Komarneni, C. H. Zhou, D. S. Tong, H. M. Yang, W. H. Yu, and H. Wang, (2016), Review Fly Ash-Based Geopolymer: Clean

Production, Properties and Applications, Journal of Cleaner Production, Vol. 125, pp. 253-267 (https://doi.org/10.1016/j.jclepro.2016.03.019).

26. T. Xie and T. Ozbakkaloglu, (2015), Behavior of Low-Calcium Fly and Bottom Ash-Based Geopolymer Concrete Cured at Ambient Temperature, Ceramics International, Vol. 41, pp. 5945-5958 (https://doi.org/10.1016/j.ceramint.2015.01.031).
27. M. Panigrahi, R. I. Ganguly, R. R. Dash, (2018), Development of Novel Constructional Material From Industrial Solid Waste as Geopolymer for Future Engineers IOP Conference Series: Materials Science and Engineering, Vol. 410, pp. 1-12.
28. M. Panigrahi, R. I. Ganguly, R. R. Dash, (2018), An Advanced Cured High Carbon Ferrochrome Slag (HCFCS) Geopolymer (GP): A Constructional Materials, IOP Conference Series: Materials Science and Engineering, Vol. 410 (1), pp. 1-15.
29. R.H. Abdul Rahim, T. Rahmiati, K. A. Azizli, Z. Man, M.F. Nuruddin, and L. Ismail (2015), Comparison of using NaOH and KOH Activated Fly Ash-based Geopolymer on the Mechanical Properties, Materials Science Forum, Vol. 803, pp 179-184 (doi:10.4028/www.scientific.net/MSF.803.179).
30. H. Öz, N. Doğan-Sağlamtimur, A. Bilgil, A. Tamer, and K. Günaydin, (2021), Process Development of Fly Ash-Based Geopolymer Mortars in View of the Mechanical Characteristics, Materials (Basel). Vol. 14, 2935 (doi: 10.3390/ma14112935).
31. I. Luhar, and S.Luhar, (2022), A Comprehensive Review on Fly Ash-Based Geopolymer, Journal of Composites Science, Vol. 6, pp. 219-59. (https://doi.org/ 10.3390/jcs6080219).
32. W. Yang, P. Zhu, H. Liu, X. Wang, W. Ge, and M. Hua, (2021), Resistance to Sulfuric Acid Corrosion of Geopolymer Concrete Based on Different Binding Materials and Alkali Concentrations, Materials, Vol. 14, pp. 7109-15 (https://doi.org/10.3390/ma14237109).
33. T. Lingyu , H. Dongpo, Z. Jianing, and W. Hongguang, (2021), Durability of Geopolymers and Geopolymer Concretes: a Review, Reviews on Advanced Materials Science, Vol. 60, pp. 1–14 (https://doi.org/10.1515/rams-2021-0002).
34. A. S. Tigue, R.A. J. Malenab, J. R. Dungca, D. E. C. Yu, and M. A. B. Promentilla, (2018), Chemical Stability and Leaching Behavior of One-Part Geopolymer from Soil and Coal Fly Ash Mixtures, Minerals, 8(9):411-19 (DOI: 10.3390/min8090411).
35. W. Yang,[1] P. Zhu, H. Liu, X. Wang, W. Ge, and M. Hua, (2021), Resistance to Sulfuric Acid Corrosion of Geopolymer Concrete Based on Different Binding Materials and Alkali Concentrations, Materials (Basel). Vol. 14, pp. 7109-15 (doi: 10.3390/ma14237109).
36. A.A. Patil, H.S. Chore, and P.A. Dodeb. (2014), Effect of Curing Condition on Strength of Geopolymer Concrete. Advances in Concrete Construction, Vol. 2, pp. 29-37.

37. M.F. Nurruddin, S. Haruna, B.S. Mohammed, and I.G. Sha'aban, (2018), Methods of Curing Geopolymer Concrete: A Review, International Journal of Advances in Applied Sciences, Vol. 5, pp. 31-36.
38. M. K. Panigrahi, (2021), Investigation of Structural, Morphological, Resistivity of Novel Electrical Insulator: Industrial Wastes, Bulletin of Scientific Research, Vol. 3, pp. 51-58.
39. M. K. Panigrahi, (2022), Investigation of Structures of Sintered Fly Ash Materials: Resources of Industrial Wastes, Bulletin of Scientific Research, Vol. 4, pp. 1-10.
40. C.N. Harman and N.S.S. Rama Rao. (2013), A Process for the Recovery of Chromium and Iron Oxide in high Carbon Ferro Chrome Slag to Obtain Chromium and Iron in the form of Saleable Metal. The thirteenth International Ferroalloys Congress Efficient technologies in ferroalloy industry, Almaty, Kazakhstan. pp. 103-108.
41. H. Wanga, M.M. Nzottab, L. Tengc, and S. Seetharamanc. (2013), Decarburization of Ferrochrome and High Alloy Steels with Optimized Gas and Slag Phases towards Improved Cr Retention, Journal of Mining and Metallurgy, Section B: Metallurgy, Vol. 49, B, pp. 175-181.
42. M. Erdem, H.S. Altundogan, M.D. Turan, and F. Tumen. (2005), Hexavalent Chromium Removal by Ferrochromium Slag, Journal of Hazardous Materials B, Vol. 126, pp. 176-182.
43. Z. Zhang, J.L. Provis, A. Reid, and H. Wang. (2014), Review Geopolymer Foam Concrete: An Emerging Material for Sustainable Construction. Construction and Building Materials, Vol. 56, pp. 113-127.
44. M.I. Khan, K. Azizli, S. Sufian, A.S. Khan, H. Ullah, and Z. Man. (2014), Geopolymer as Sustainable Binder of the 21st Century: A Review, World Sustainability Forum, 2014-Conference Proceedings Paper.
45. V. Kumar, V.K. Singh, and A. Srivastava. (2013), Low Temperature Synthesis of High Alumina Cements by Novel Co-Melt Precursors and Their Implementation as Castables with Some Micro Fine Additives, Journal of the American Ceramic Society, Vol. 96, pp. 2124-2131.
46. P.H. Kumar, A. Srivastava, V.Kumar, M.R. Majhi, and V.K. Singh. (2014), Implementation of Industrial Waste Ferrochrome Slag in Conventional and Low Cement Castables: Effect of Micro Silica Addition, J. Asian Ceramic Society, Vol. 2, pp. 169-175.
47. Z. Cao, L. Xiang, E. Peng, and K. Li, (2018), Experimental Study on Electrical Resistivity of Cement-Stabilized Lead-Contaminated Soils, Geomaterials in Geotechnical Engineering, Vol. 2018, pp. xxx-xxx (https://doi.org/10.1155/2018/4628784).
48. F. Farhana, H. Kamarudin, A. Rahmat, Mohd M. Al Bakri Abdullah, and S. R. Shamsudin, (2015), The Electrical Resistivity of Geopolymer Paste by Using Wenner Four Probe Method, Key Engineering Materials, Vol. 660: pp. 28-33 (DOI: 10.4028/www.scientific.net/KEM.660.28).

49. Jingming Cai, Jinlong Pan, Xiaopeng Li, Jiawei Tan, and Jiabin Li, (2020), Electrical resistivity of fly ash and metakaolin based geopolymers, Construction and Building Materials, Vol. 234(2), pp. 117868-xxx (DOI: 10.1016/j.conbuildmat.2019.117868).
50. Walter J, Uthayakumar M, Balamurugan P, Mierzwiński D. (2021), The Variable Frequency Conductivity of Geopolymers during the Long Agieng Period. Materials (Basel). Vol. 14, pp. 5648-17 (doi: 10.3390/ma14195648).
51. S.D. Muduli, J.K. Sarangi, B.D. Nayak, B.K. Mishra. (2013), Effect of NaOH Concentration in Manufacture of Geopolymer Fly Ash Building Brick. Greener Journal of Physical Sciences, Vol. 3, pp. 204-211.
52. J.D. Russell, (1987), Infrared methods. In: A Hand Book of Determinative Methods in Clay Mineralogy. Wilson MJ (Ed.) Blackie and Son Ltd, NY, Vol. 133, pp. 11-67.
53. K. Ramaswamy and R. Venkatachalapathy, (1992), Infrared Spectroscopic Analysis of Sedimentary Formations of Neyveli Lignite Minecut-Ii, Indian Journal of Pure & Applied Physics, Vol. 30, pp. 171-xxx.
54. G. D. Poggetto, A. D. Angelo, I. Blanco, S. Piccolella, C. Leonelli, and M. Catauro, (2021), FT-IR Study, Thermal Analysis, and Evaluation of the Antibacterial Activity of a MK-Geopolymer Mortar Using Glass Waste as Fine Aggregate,Polymers, Vol. 13, pp. 2970-xxx (https://doi.org/10.3390/polym13172970).
55. C. Finocchiaro, G. Barone, P. Mazzoleni, C. Leonelli, A. Gharzouni, S. Rossignol, (2020), FT-IR Study of Early Stages of Alkali Activated Materials based on Pyroclastic Deposits (Mt. Etna, Sicily, Italy) using two different alkaline solutions, Construction and Building Materials, Vol. 262, pp.120095-xxx (https://doi.org/10.1016/j.conbuildmat.2020.120095).
56. W. K. W. Lee and J. S. J. van Deventer, (2003), Use of Infrared Spectroscopy to Study Geopolymerization of Heterogeneous Amorphous Aluminosilicates, Langmuir , Vol. 19, pp. 8726–8734 (https://doi.org/10.1021/la026127e)
57. M. Panigrahi, S. Mohanty, R. I. Ganguly, R. R. Dash "Synthesis and Characterizations of High Carbon Ferrochrome (HCFC) Slag Based Geopolymer, IntechOpen Publisher, UK.ISBN: 978-1-83968-989-5, (DOI: http://dx.doi.org/10.5772/intechopen.97140).

7

High Resistant PA/HCFC Slag Geopolymer Materials

Muktikanta Panigrahi[1]*, Ratan Indu Ganguly[2] and Radha Raman Dash[3]

[1]Department of Materials Science, Maharaja Sriram Chandra Bhanja Deo University, Keonjhar Campus, Odisha, India
[2]Department of Metallurgical Engineering, National Institute of Technology, Rourkela, Odisha, India
[3]CSIR-National Metallurgical Laboratory (NML), Jamshedpur, Jharkhand, India

Abstract
The objective of this chapter is to explore the possibilities of using pond ash/high-carbon ferrochrome (PA/HCFC) slag geopolymer (GP) for developing insulators. The electrical property of PA/HCFC slag GP is also studied.

Pond ash (PA) is a waste material of thermal power stations with low-calcium content. It has high silica (62.8%) and alumina (28.3%). High-carbon ferrochrome (HCFC) slag is a by-product of the ferrochrome industry. It contains silica (28%) and alumina (23.5–24.5%). It is thought to be an important proposition to balance alumina and silica ratio by mixing the above two materials judicially. This will help in the formation of GP with good strength. Researchers have innovatively established the constructive use of geopolymer in modern construction work. The PA and HCFC slag are mixed in requisite proportion and are treated with alkali materials. Thereafter, they are cured at 70 °C for 24 h to manufacture PA/HCFC slag GP, which is characterized by XRD, SEM with EDS, TEM, FTIR. Compressive strength, Vickers hardness, and electrical resistivity are measured. A corrosion study is made to evaluate corrosion property, and thermal characteristics of PA/HCFC slag GP material are studied by TGA and DSC. The SEM analysis of PA/HCFC slag GP shows crystalline and non-crystalline phases. Compressive strength of PA/HCFC slag-based GP is determined to be 22 MPa, and resistivity of PA/HCFC slag GP is estimated to be 125.79705×10^6 (Ohm.cm). The formation of geopolymer was indicated by DSC isotherm analysis, and TGA analysis was

**Corresponding author*: muktikanta2@gmail.com

Muktikanta Panigrahi, Ratan Indu Ganguly and Radha Raman Dash (eds.) High Electrical Resistant Materials: Ferrochrome Slag Resource Ceramics, (159–176) © 2024 Scrivener Publishing LLC

performed to study the stability of prepared GP products. Corrosion resistance of the material is highly encouraged in normal water.

Keywords: Pond ash, HCFC slag, geopolymer, microscopic analyses, strength properties, TGA, DSC, resistivity, corrosion study

7.1 Introduction

Important technological developments in recent decades have promised production of novel materials such as geopolymer (i.e., alkali-activated alumino-silicates) [1–4]. This material has attracted the attention of the scientific community, particularly civil/structural engineers. Advancements in the use of this material are encouraging. However, this material has not yet been fully exploited by civil/structural engineers. This material has unique properties, i.e., low carbon footprint, excellent mechanical properties and high corrosion resistance (heat and acids) [1, 4–8]. Moreover, production of this material is cost-effective and safeguards the Earth's environment.

It is prudent to analyze raw materials scientifically before proceeding to develop a process for forming GP. For this purpose, XRD and SEM image analyses of raw materials are considered important.

Pond ash is a by-product of thermal power stations which contains silica (62.8%) and alumina (28.3%) [9–11]. However, this material possesses low calcium.

Ferrochrome (FeCr) slag is a waste material which is formed during carbothermic reduction of FeCr ore. It is also estimated that globally, annual production of FeCr alloy is between 12 and 16 Mt [12, 13]. It is estimated that per ton ferroalloy production, 1.1–1.5 tons of slag is produced [12, 13]. FeCr slag is mostly disposed of in landfills and this creates environmental problems [12, 13].

Since it contains an elevated level of toxic elements, FeCr slag is a well-known pollutant [14, 15]. Potential release of these elements, especially Cr(VI), limits its use [16]. Exposure to Cr(VI) has been shown to cause lung cancer, pulmonary and dermatological problems [16]. In a country like Finland, FeCr slag is classified as a non-hazardous waste [17]. However, in South Africa, FeCr is classified as a hazardous waste [18]. It is therefore imperative that more research be carried out to explore where FeCr is used along with other components to address health hazard issues. By utilizing FeCr slag with other waste materials, there will be a reduction of the tonnage of slags accumulating in landfills, thereby fulfilling the demands of sustainable development through the use of secondary resources in the construction industry.

There are different ways by which FeCr slag can be used. One way is to use Portland cement, which acts as a binder [19]. Although cement is readily available, it has been shown that Cr hinders hydration of cement, leading to long-term leachability problems in the hardened structures [18, 19]. FeCr slag has also been used in construction of flexible pavements. This aggregate is reported to have minimal environmental contamination [20, 21]. The FeCr slag is mixed with zeolites to produce bricks, which can be used for construction purposes. However, there is not more data available to determine metal leachability of the bricks [22]. By stabilizing FeCr slag with the alkaline activation method, the leaching problem can be minimized. The resultant product formed is known as geopolymer (GP). Geopolymer is a three-dimensional amorphous or semi-crystalline inorganic polymer [23]. It is durable, acid resistant and has high compressive strength [24]. Thus, it is very much possible to immobilize toxic and nuclear wastes by synthesizing GP with the alkali activation method [25].

References are available on resistivity measurement of geopolymer paste. Alkali activation of FeCr slag [26, 27] results in formation of geopolymer (cured 7 days), which yields maximum unconfined compressive strength (UCS) of 20.8 MPa. However, UCS is reported to have dropped from 20.8 MPa to 13.1 MPa. These studies have not been considered for the immobilization of metals during alkaline activation but considered for material stability in an environment of fire and sulphate resistance. The environment acts as an important barrier for usage of FeCr slag in the construction industry.

Kljajević *et.al.* [28] synthesized geopolymers (GP) from metabentonite (MB), metadiatomite (MD) and metakaolinite (MK). Different test methods such as XRD, FTIR, and SEM/EDX are adopted to characterize the synthesized GP. Geopolymers are quasi-solid-state electrolytes and possess a high electrical conductivity. The highest conductivity observed for GP with metakaolin (GPMK) is 2.14×10^{3} Ω.cm at 700 °C. The corresponding activation energies of conductivity for this sample amounted to 0.33 eV in the temperature range of 500–700 °C. The geopolymer synthesized from metakaolin has good ionic conductivity values, which recommends its use as an alternative material for a solid oxide fuel cell (SOFC).

Farhana *et al.* [29] presented electrical resistivity of geopolymer paste using Wenner four-probe methods. Rectangle sized cured GP is prepared from aluminosilicate materials, i.e., fly ash and an alkaline activator. Geopolymer paste (after 28 days, 0.01 mA current and 0.10 m electrode spacing) shows the highest resistivity (61575 Ωm), whereas geopolymer paste (after 7 days, 0.95 mA current and 0.02 m electrode spacing) shows the lowest resistivity (537 Ωm). Thus, corrosion rate of geopolymer paste is very low (i.e., negligible).

Polder [30] described how to assess on-site concrete resistivity problems related to corrosion and protection of reinforcement. Electrical resistivity of concrete can be related to the two processes involved in corrosion of reinforcement: initiation (chloride penetration) and propagation (corrosion rate). The resistivity of a structure exposed to chloride indicates the risk of early corrosion damage, because a low resistivity is related to rapid chloride penetration and to a high corrosion rate. Once damage has occurred, resistivity is relevant for electrochemical maintenance methods. The resistivity of the cover (near surface) concrete can be measured non-destructively by placing electrodes on the concrete surface, applying a voltage and measuring the current. Several arrangements can be used, one electrode (the reinforcement is the second electrode), two electrodes, and four electrodes. Practical aspects of these arrangements are discussed. Procedures for calibration are proposed and guidelines for the interpretation and reference values for various exposures and concrete compositions are given.

Song and Kwon [31] prepared reinforced concrete structures. Concrete structures have potential to withstand different environmental conditions. However, failures in the structures do still occur as a result of premature reinforcement corrosion. The maintenance and repair of bridges/buildings (for their safety) need proper inspection via monitoring techniques to assess reinforcement corrosion. The electrochemical method is used to assess corrosion, and is also applied for inspecting bridges, buildings and other civil engineering structures.

Glass *et al.* [32] protected steel in concrete through conductive coatings using cathodic protection method. Property of such coating is evaluated as low sheet resistance (square resistance). Current density and anode connection geometry determines uniformity of current distribution over concrete surface. Sheet resistance is estimated by four-probe technique using (p/ln2)6(DV/I), which is a non-destructive testing.

In the present work, high-carbon ferrochrome slags (HCFC slag) and pond ash (PA) are used as raw materials. They are activated by sodium hydroxide (NaOH) solution having different concentrations during the preparation of geopolymer. Combined effects of PA and HCFC slag treated with activated alkaline solution on the geopolymer properties are studied. Different phases are identified by X-ray diffraction pattern. FTIR spectra are indicated to determine different absorption bands. TG-DTG is studied to explore the stability and DSC experiment is used to determine the condensation polymerization. Surface topography is shown by SEM.

7.2 Experiment Details

7.2.1 Source of Materials

Pond ash is obtained from the National Aluminium Company (NALCO), a Navratna Central Public Sector Enterprise (Damanjodi, Odisha, India). HCFC slag (main ingredient for geopolymer preparation) is procured from Shyam Ferro Alloys (Durgapur, West Bengal, India). Sodium hydroxide (NaOH) is used as an activator. This was purchased from Loba Chemie, India and sodium silicate (Na_2SiO_3) is procured from Merck, India. Water-soluble plasticizer (Sika) was purchased from a reputed firm located in the southern part of India (Visakhapatnam market).

7.2.2 PA/HCFC Slag Geopolymer (GP) Preparation

Steps for preparing PA/HCFC slag-based GP are described below [33, 34].

Step 1: Raw material preparation
Geopolymer is prepared from pond ash, HCFC slag, sodium silicate, and alkali. HCFC slag is crushed and ground to 240 mesh size. Mesh particles are then dried in an oven for 2 h at 120 °C.

Step 2: Geopolymer preparation with raw materials
The ground and dried materials are mixed with requisite amount of NaOH (8 M), Na_2SiO_3 and water-soluble plasticizer (Sika) (Table 7.1).

Step: Molding, casting and compaction
Semi-solid mass is placed in a cylindrical mold (55 mm diameter and 70 mm height). The mixture is rammed in the mold 15 times to ensure

Table 7.1 PA/HCFC slag GP (three samples with same compositions).

Sample code	PA (51%) + HCFC slag (34%)	Sod. Silicate (SS)	Alkali (8 M, NaOH) (SH)	Water-soluble plasticizer (Sika)
S1	85%	12%	3%	1-2 ml
S2	85%	12%	3%	1-2 ml
S3	85%	12%	3%	1-2 ml

uniformity of mixture. Before demolding, ramming mixture is allowed to be kept for 3 minutes in the mold. This enables compaction of mass and is known as green sample.

Step 4: Curing process
Green samples are cured at a temperature of 70 °C for 24 h in a heating oven. Prepared cured samples are then characterized by XRD, FTIR TGA, DSC, etc.

7.2.3 Test Methods

X-ray diffraction (XRD) analyses are used to identify different phases. Philips PW-1710 Advance wide-angle X-ray diffractometer and Philips PW-1729 X-ray generator are used for X-ray analyses. XRD machine functions at 40 kV and 20 mA. Prepared pellet sample is directly placed on sample holder (made from quartz). Sample is scanned through diffraction angle (2θ) varying between 10° to 70° with scanning rate of 2°/min.

Morphological analyses are carried out with a Carl Zeiss Supra-40 microscope. Samples are gold-coated using sputtering technique. An instrument parameter, i.e., operating voltage is kept at 30 kV, is used during testing. Energy dispersive X-ray spectroscopy (EDS) is performed along with microscopic examination by SEM.

A transmission electron microscope (5022/22, Tecnai G220 S-Twin, Czech Republic)) is used to investigate minute details of images. Selected area diffraction (SAD) pattern of fused spots from microstructure of raw HCFC slag powder and PA/HCFC slag GP are analyzed. TEM specimen (HCFC slag) is prepared by drop casting technique. TEM specimen of PA/HCFC slag GP is prepared by microtone technique (Leica Microsystems GmbH, A-1170). Then, specimens are transferred to carbon-coated Cu TEM grids.

The FTIR spectra of raw HCFC slag and PA/HCFC slag GP are obtained in a Thermo Nicolet Nexus 870 spectrophotometer (range 400–4000 cm^{-1}). Settings parameters (i.e., 50 scan at 4 cm^{-1} resolution, absorbance measurement mode) are kept constant. Pellet sample is prepared with bulk amount of KBr powder. Test samples are prepared by compression molding system in a pelletizer. A background spectrum is collected before the test. Then, pellet is put into a FTIR sample holder and data are collected.

In electrical resistivity measurement, test samples in the pellet form are used. Resistivity is estimated using the following equation [35, 36].

$$\rho = \left(\frac{R \times A}{l}\right) \tag{7.1}$$

Here, A is the area of electrode $= \frac{\pi}{4} D^2$ (cm^2), R is the resistance of prepared samples (Ω), l is thickness of the sample, and ρ is the resistivity of the samples in Ω.cm. For this measurement, a Keithley source/multimeter is used.

7.2.4 Results and Discussion

Table 7.2 shows compressive strength values for three different materials, viz., PA/HCFC Slag GP, HCFC Slag GP, and PA GP materials. All three materials are prepared by the same treatment combination. The maximum strength is observed for PA/HCFC Slag GP, which is 22 MPa.

The other two materials, namely HCFC Slag GP and PA GP, have shown strength value 16 MPa and 19 MPa, respectively. This is equivalent to the standard M20 grade concrete used for making structure [33, 34].

Figure 7.1a-b shows microstructure of materials, i.e., as-received HCFC slag and as-received pond ash (PA). Both images show the presence of crystalline phase and some areas containing glassy phase. Figure 7.1c-d designate the microstructure of geopolymer prepared from HCFC slag and PA at lower and higher magnification, respectively. It is seen from these images (Figure 7.1a-b) that particles have combined in many areas due to polymerization reaction occurring during treatment.

Table 7.2 Compressive strength of PA/HCFC slag GP materials [37].

S.no.	Sample type with combinations	Optimal parameters	Av compressive strength (MPa)
1	PA/HCFC Slag GP (PA+HCFC slag=85%)	24 h 70 °C oven curing; 12% 8M NaOH, 3% Na_2SiO_3, 1-2 mL Sika	22
2	HCFC Slag GP (HCFC slag=85%)	24 h 70 °C oven curing; 12% 8M NaOH, 3% Na_2SiO_3, 1-2 mL Sika	16
3	PA GP (PA=85%)	24 h 70 °C oven curing; 12% 8M NaOH, 3% Na_2SiO_3, 1-2 mL Sika	19

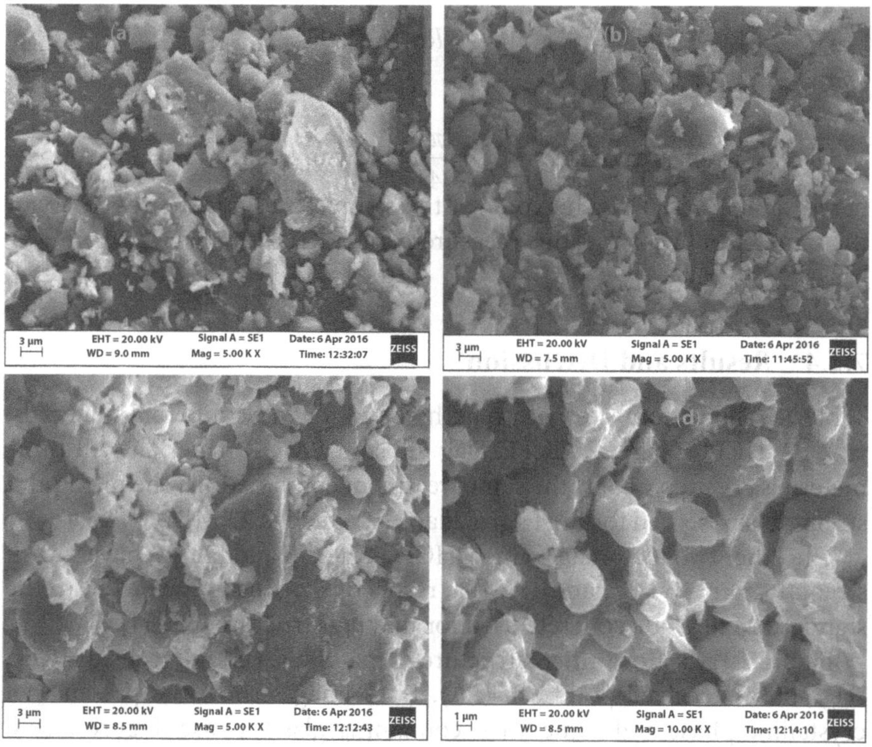

Figure 7.1 FESEM image of HCFC slag (a), PA (b), PA/HCFC Slag GP with lower magnification (c), and PA/HCFC Slag GP with higher magnification (d).

Energy dispersive X-ray (EDX) microanalysis (Figure 7.2) confirms presence of elements such as aluminium, silicon, and oxygen which support formation of Geopolymer with Si, Al, Fe, Mg, Ca, Cr, Na, Ti, and O, etc.

In order to resolve particles with resolution, HRTEM analyses have been done for raw pond ash. Figure 7.3a, 7.3b, and 7.3c show microstructures of raw pond ash at different magnification. Figure 7.3a shows a few particles which are spread over the background having different sizes. Shapes of particles are either spherical or elliptical. Aspect ratios of particles vary between 1.0 to 1.5.

Upon magnification (i.e., 500 nm) with high resolution within a particle (Figure 7.3b), it is observed that there are other small particles. If it is further magnified (Figure 7.3c), embedded particles are seen to be rod-shaped. However, resolution decreases with higher magnification.

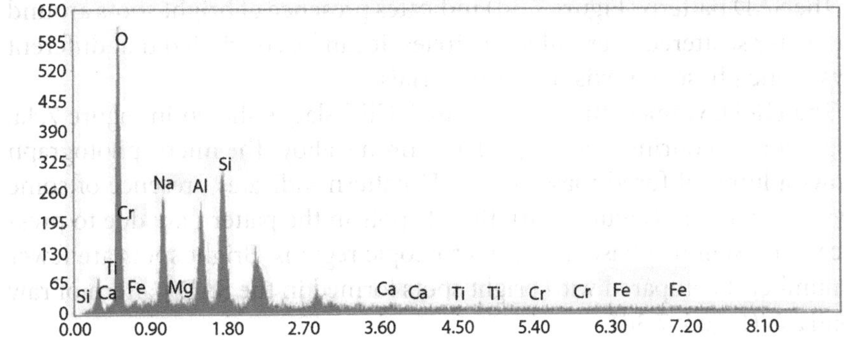

Figure 7.2 The EDX spectrum of PA HCFC slag-based GP.

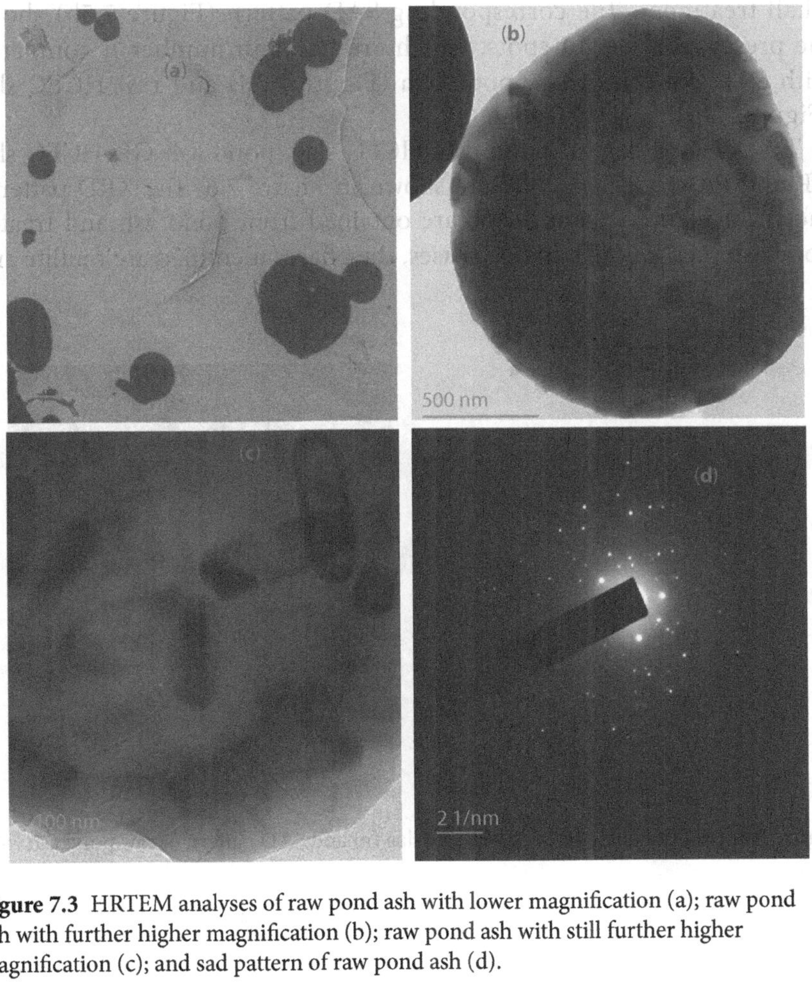

Figure 7.3 HRTEM analyses of raw pond ash with lower magnification (a); raw pond ash with further higher magnification (b); raw pond ash with still further higher magnification (c); and sad pattern of raw pond ash (d).

The SAD pattern (Figure 7.3d) indicates presence of bright spots around the center scattered over different circles. It can be concluded that different crystalline phases co-exist in PA materials.

The HRTEM microstructure of raw HCFC slag is shown in Figure 7.4a. Slag is formed during smelting of ferrochrome alloy. The micro-photograph shows a lump of fused mass. The SAD pattern indicates presence of some crystalline phases (Figure 7.4b). Bright spots in the pattern are due to presence of crystalline phase in the microscopic regions. Bright spots are fewer in number if compared with bright spots formed in the SAD pattern of raw pond ash (Figure 7.3d).

The HRTEM micrograph of PA/HCFC slag geopolymer is shown in Figure 7.5a. It is seen that there is coagulation of particles, which is due to the geopolymerization process occurring in the materials during alkali treatment. The corresponding SAD pattern (Figure 7.5b) shows the presence of bright spots, which are fewer in number if compared with SAD pattern of raw pond ash (Figure 7.3d) and raw HCFC slag (Figure 7.4b).

The XRD pattern of pond ash, HCFC slag, pond ash GP, HCFC slag GP, and PA/ HCFC slag GP are shown in Figure 7.6. The XRD patterns shown in Figure 7.6a and 7.6c are obtained from pond ash and treated pond ash (i.e., PA GP). In both cases, the phases identified are mullite and

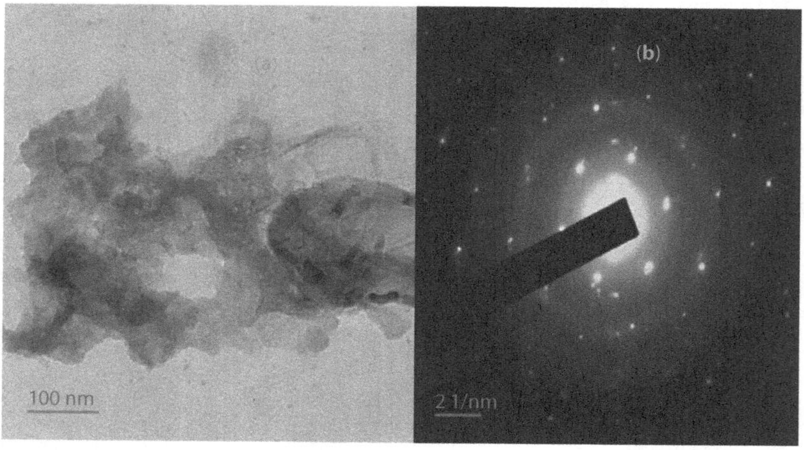

Figure 7.4 HRTEM analyses of raw HCFC slag (a) and SAD pattern of raw pond ash (b).

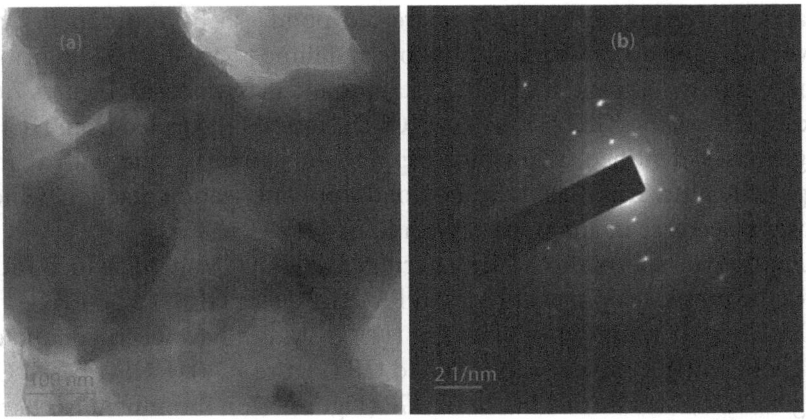

Figure 7.5 HRTEM analyses of PA/HCFC slag GP (a) and SAD pattern of PA/HCFC slag GP (b).

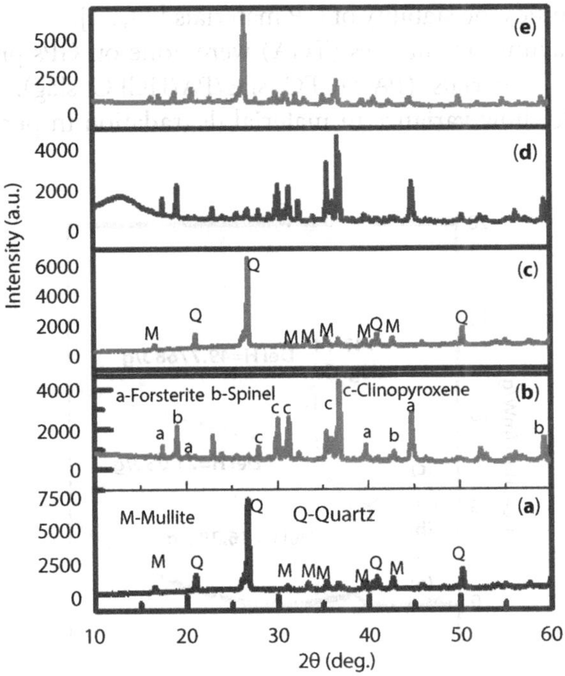

Figure 7.6 The XRD pattern of PA (a), HCFC Slag (b), PA GP (c), HCFC slag GP (d), and PA-HCFC slag GP (e).

quartz. However, peak heights of treated pond ash have dropped. This advocates for a decrease in quantity of crystalline phase due to GP formation. Geopolymer contains both crystalline and glassy phases. Figures 7.6b and 7.6d show the XRD patterns of HCFC slag and GP formed from HCFC slag. Phases identified are forsterite, spinel, and clinopyroxene, with a drop in peak heights. This is an indication of the polymerization reaction, which causes a decrease in amount of crystalline phase.

Figure 7.6e shows the XRD pattern of PA/HCFC slag GP, which indicates the presence of many phases in PA/HCFC slag GP.

Differential scanning calorimetry (DSC) studies have been conducted on geopolymer samples, i.e., PA-based GP (Figure 7.7a), HCFC-based GP (Figure 7.7b), and PA/HCFC-based GP (Figure 7.7c). Prominent isotherm peaks are observed for all the above materials at around 100 °C (PA GP) and 150 °C (HCFC GP and PA/HCFC GP). Endothermic peak has shifted to a higher temperature for HCFC GP and PA/HCFC GP. However, there is a large change in delta H values, which is due to addition of PA to HCFC slag. This signifies the stability of GP materials [33, 34].

Thermogravimetric analyses (TGA) were done on GPs prepared from different raw materials (PA/HCFC slag/PA/HCFC slag). Figure 7.8a, 7.8b, and 7.8c show variance in material degradation in percentage with

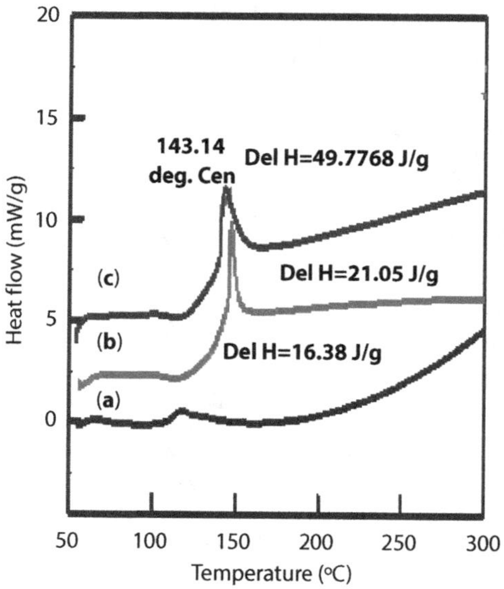

Figure 7.7 DSC isotherm of PA GP (a), HCFC GP (b), and PA/HCFC GP (c) [37].

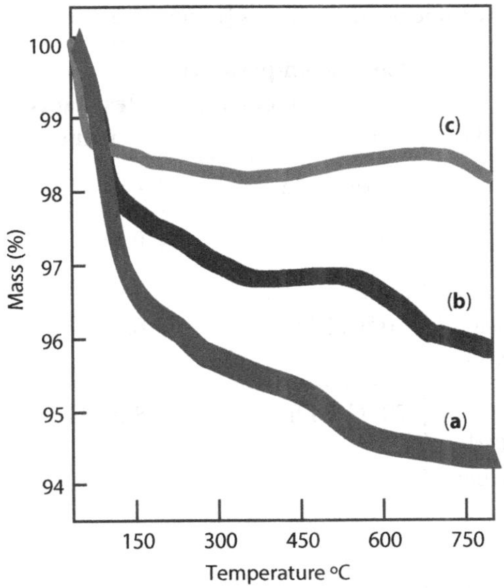

Figure 7.8 TGA curve of PA GP (a), HCFC slag GP (b), and PA/HCFC slag-based GP (c) [37].

temperature. Weight loss (%) with increasing temperature shows relative stability of GP material at higher temperature. Pond ash geopolymer drops in weight by 4% at 120 °C. Whereas, HCFC slag GP and PA/HCFC slag GP have weight loss of 2% and 1.4%, respectively. There is continuous loss in weigh recoded for PA GP and HCFC slag GP. Initial weight loss for all three materials is due to loss of moisture from the materials beyond 120 °C. PA/HCFC slag GP shows stability beyond 150 °C and has maintained stability till 750 °C. HCFC slag GP has shown intermediate behavior between PA GP and PA/HCFC slag GP.

Resistance values of PA GP, HCFC slag GP, and PA/HCFC slag GP are determined by two-probe method (see Table 7.3). Estimated resistivity values of PA GP, HCFC slag GP, and PA/HCFC slag GP are shown in Table 7.3. The highest resistivity value is found to be 125.79705×10^9 Ohm.cm for HCFC slag/PA GP, which is cured at 70 °C and 24 h. The lowest resistivity value is estimated to be 10.74222×10^6 Ohm.cm for PA GP. Difference in resistivity value is due to the compositional difference of geopolymer formed from three different materials [38–41].

Table 7.3 Resistivity value of GP materials [radius=1.5 cm and thickness=0.5 cm].

S.no.	Sample ID	Curing temperature (°C) and curing time	Resistance value	Resistivity value (Ω.cm)
1	Pond Ash	As received	---	---
2	Pond Ash Geopolymer	70 °C, 24 h	0.76 MΩ	10.74222×10^6
3	HCFC slag Geopolymer	70 °C, 24 h	0.9 GΩ	12.72×10^9
4	Pond Ash/ HCFC slag Geopolymer	70 °C, 24 h	8.9 GΩ	125.79705×10^9

7.3 Conclusions

Pond ash/high-carbon ferrochrome slag (PA/HCFC)-based geopolymer (GP) was prepared with optimum treatment combination and curing condition. Compressive strength of PA/HCFC slag-based GP was determined to be 22MPa.

In two-probe methods, electrical resistivity of as-prepared PA/HCFC slag GP is measured to be 125.79705×10^9 (Ohm.cm).

Based on SEM analysis, there were marked differences in microstructure of three GP materials. PA/HCFC slag GP shows both crystalline and glassy phases in the matrix. In TEM studies, PA/HCFC slag GP materials showed fused mass, whereas individual raw materials showed embedded irregular spherical particles (for PA) and irregular shaped hexagonal crystals (for HCCF slag). The presence of desired elements in PA/HCFC slag GP was proven by EDS studies. DSC and TGA provided well-supported data for studying thermal behavior of as-prepared GP materials.

In view of compressive strength data, structure integrity, thermal results, and corrosion data, it is evident that PA/HCFC slag GP is suitable material for high-strength structural applications. Because of its good mechanical properties, it is expected to make building products, i.e., brick and concrete, which will be more sustainable in terms of environmental performance, economic viability and potential for waste utilization.

Acknowledgments

The authors convey their sincere thanks to the Ministry of Mines, Government of India, for providing the grant to carry out the work [Grant number= F.No.:14/54/20214-Met.-IV; dated: 29.12.2014]. They also convey their sincere thanks to GIET, University Gunupur, Rayagada, Odisha, India for providing lab facilities for the research work. They would also like to thank CRF IIT Kharagpur for providing testing facilities. The authors would like to thank Prof. Munesh Chandra Adhikary, PG Council Chairman, Fokir Mohan University, for his invaluable guidance, advice, and constant inspiration throughout the entire process. They would also like to thank Mr. Mukteswar Mohapatra of Fokir Mohan University for his support.

References

1. P. Benito, C. Leonelli, V. Medri, and A. Vaccari, (2013), Geopolymers: a New And Smart Way for A Sustainable Development, Applied Clay Science, Vol. 73, p. 1-xxx.
2. H.A. Abdel Gawwad, S. Abd El-Aleem, and A.S. Ouda, (2016), Preparation and Characterization of One-Part Non-Portland Cement, Ceramics International, Vol. 42, pp. 220-228.
3. W. Zhu, X. Chen, A. Zhao, L. J. Struble, and E.-H. Yang, (2019), Synthesis of High Strength Binders from Alkali Activation of Glass Materials from Municipal Solid Waste Incineration Bottom Ash, Journal of Cleaner Production, Vol. 212, pp. 261-269.
4. L. Yang, E. Yilmaz, J. Li, H. Liu, and H. Jiang, (2018), Effect of Superplasticizer Type and Dosage on Fluidity and Strength Behavior of Cemented Tailings Backfill with Different Solid, journal name, and Vol.xxx, pp. 290-298.
5. J.-P. Qiu, L. Yang, J. Xing, and X.-G. Sun, (2018), Analytical Solution for Determining the Required Strength of Mine Backfill based on Its Damage Constitutive Model, Soil Mechanics and Foundation Engineering, Vol. 54, pp. 371-376.
6. F. Slaty, H. Khoury, J. Wastiels, and H. Rahier, (2013), Characterization of Alkali Activated Kaolinitic Clay, Applied Clay Science, Vol. 75-76, pp. 120-125.
7. I. Ozer and S. Soyer-Uzun, (2015), Relations Between the Structural Characteristics and Compressive Strength in Metakaolin Based Geopolymers with Different Molar Si/Al Ratios, Ceramics International, Vol. 41, pp. 10192-10198.

8. L. Yang, J. Qiu, H. Jiang, S. Hu, H. Li, and S. Li, (2017), Use of Cemented Super-Fine Unclassified Tailings Backfill for Control of Subsidence, Minerals, Vol. 7, p. 216-xxx.
9. A.K. Bera, A. Ghosh, and A. Ghosh, (2017), Compaction Characteristics of Pond Ash, Journal Of Materials In Civil Engineering, Vol. 19, pp. 349-357 (DOI: 10.1061/(ASCE)0899-1561 (2007)19:4(349)).
10. K. Arumugam, R. Ilangovan, and D.M. James, (2011), A Study On Characterization And Use of Pond Ash as Fine Aggregate in Concrete, International Journal Of Civil And Structural Engineering Vol. 2, pp. 466-474.
11. J. Mallick and D.P. Palai, (2016), Collapsible Behavior of Pond ash, International Research Journal of Engineering and Technology (IRJET), Vol. 3, pp. 480-485.
12. B. B. Lind, A.-M. Fallman, and L. B. Larsson, (2001), Environmental Impacts of Ferrochrome Slag in Road Construction, Waste Management, Vol. 21, pp.255-264.
13. W. Shao-Peng, Y. Wen-Feng, X. Yong-Jie, and L. Zhen-Hua. (2003), Design and Preparation of Steel Slag. J. Wuhan University of Technology, Vol.18, pp. xxx-xxx.
14. W. Shao-Peng, Y. Wen-Feng, X. Yong-Jie, and L. Zhen-Hua. (2003), Design and Preparation of Steel Slag. J. Wuhan University of Technology, Vol.18, pp. xxx-xxx.
15. M. Tossavainen, (2018), Leaching Results in the Assessment of Slag and Rock Materials as Construction Material. 2005. (http://epubl.ltu.se/1402-1544/2005/44/LTU-DT-0544-SE.pdf-accessed on 31 August
16. P. Sharma, S. P. Singh, S. K. Parakh, and Y. W. Tong, (2022), Health hazards of hexavalent chromium (Cr (VI)) and its microbial reduction. Bioengineered. Vol. 13, pp. 4923-4938 (doi: 10.1080/21655979.2022.2037273).
17. T. Falayi, (2019), Sustainable solidification of ferrochrome slag through geo-polymerisation: a look at the effect of curing time, type of activator and liquid solid ratio, Sustain Environ Res, Vol.29, pp. 1-10 (https://doi.org/10.1186/s42834-019-0022-7).
18. J. P. Beukes, P. G. Van Zyl, and M. Ras, (2012), Treatment of Cr(VI)-containing wastes in the South African ferrochrome industry - a review of currently applied methods, Journal of the Southern African Institute of Mining and Metallurgy, Vol. 112, pp. 347-352.
19. M. L. Nehdi, and A. Yassine (2020), Mitigating Portland Cement CO_2 Emissions Using Alkali-Activated Materials: System Dynamics Model, Materials (Basel), Vol. 13, pp. 4685- 23. (doi: 10.3390/ma13204685).
20. C. Ajay, K. Durga Rani, (2022), Ferrochrome Slag characterization and its compatibility with Quarry dust in Flexible pavement, Materials Science IOP Conference Series: Earth and Environmental Science, Vol. 982, pp. 012008-10 (DOI:10.1088/1755-1315/982/1/012008).

21. S. S. Patil, S. S. Bachhav, and D. Y. Kshirsagar, (2016), Use of steel slag in construction of flexible Pavement, International Journal of Engineering and Innovative Technology (IJEIT), Vol. 5, pp. 24-27.
22. O. Gencel, M. Sutcu, E. Erdogmus, V. Koc, V. V. Cay, and M. S. Gok, (2013), Properties of bricks with waste ferrochromium slag and zeolite, Journal of Cleaner Production, Vol. 59, pp. 111-119 (https://doi.org/10.1016/j.jclepro.2013.06.055).
23. G J. Davidovits, (2017), Geopolymers: Ceramic-like inorganic polymers, Journal of Ceramic Science and Technology, Vol. 8, pp. 335-350 (DOI: 10.4416/JCST2017-00038).
24. M. Amran, A. Al-Fakih, S.H. Chu, R. Fediuk, S. Haruna, A. Azevedo, and N. Vatin, (2021), Show more Case study Long-term durability properties of geopolymer concrete: An in-depth review Case Studies in Construction Materials, Vol. 15, pp. e00661-xxx.
25. I. Luhar, S. Luhar, M. M. A. B. Abdullah, R. A. Razak, P. Vizureanu, A. V. Sandu, and P. D. Matasaru (2021), A State-of-the-Art Review on Innovative Geopolymer Composites, Designed for Water and Wastewater Treatment. Materials (Basel). Vol. 14, pp. 7456. (doi: 10.3390/ma14237456).
26. M. B. Karakoc, I. Turkmen, M. M. Maras, F. Kantarci, R. Demirboga, and M. U. Toprak, (2014), Mechanical properties and setting time of ferrochrome slag based geopolymer paste and mortar, Construction and Building Materials, Vol. 72, pp. 283–92.
27. S. K. Nath, (2018), Geopolymerization behavior of ferrochrome slag and fly ash blends, Construction and Building Materials, Vol. 181, pp. 487–94.
28. L. M. Kljajević, Z, Melichová, Z. Melichová, M. Stojmenović, M. Stojmenović, S. Nenadović, and S. Nenadović, (2019), Structural and electrical properties of geopolymer materials based on different precursors (kaolin, bentonite and diatomite), Macedonian Journal of Chemistry and Chemical Engineering, Vol. 38, pp. 283-292 (DOI: 10.20450/mjcce.2019.1678).
29. F. Farhana, H. Kamarudin, A. Rahmat, Mohd M. Al Bakri Abdullah, S. Rizam, S., Mohammad, T. Selimin, and A. V. Sandu, (2015), The Electrical Resistivity of Geopolymer Paste by Using Wenner Four Probe Method, Key Engineering Materials, Vol. 660, pp. 28-33 (DOI: 10.4028/www.scientific.net/KEM.660.28).
30. R. B. Polder, (2001), Test methods for onsite measurement of resistivity of concrete - a RILEM TC-154 technical recommendation, Construction and Building Materials, Vol. 15, pp. 125-131.(https://doi.org/10.1016/S0950-0618(00)00061-1)
31. H.-W. Song, and S.-J. Kwon, (2007), Permeability characteristics of carbonated concrete considering capillary pore structure, Cement and Concrete Research, Vol. 37, pp. 909-915 (https://doi.org/10.1016/j.cemconres.2007.03.011).
32. G. K. Glass, A. M. Hassanein, and N. Buenfeld, CP Criteria for Reinforced Concrete in Marine Exposure Zones, (2000), Journal of Materials in Civil Engineering, Vol. 12, pp. xxx (DOI: 10.1061/(ASCE)0899-1561(2000)12:2(164).

33. M. Panigrahi, S. Mohanty, R.R. Dash, and R.I. Ganguly, (2018), Development of Novel Constructional Material from Industrial Solid Waste as Geopolymer, IOP Conf. Series: Materials Science and Engineering, Vol. 410, pp. 1-12 (DOI:10.1088/1757-899X/410/1/012012).
34. M. Panigrahi, S. Mohanty, R.I. Ganguly, and R.R. Dash, (2021), Synthesis and Characterizations of High Carbon Ferrochrome (HCFC) Slag Based Geopolymer, IntechOpen Publisher, UK, pp-1-16, SBN: 978-1-83968-989-5 (DOI: http://dx.doi.org/10.5772/intechopen.97140).
35. M. K. Panigrahi, (2021), Investigation of Structural, Morphological, Resistivity of Novel Electrical Insulator: Industrial Wastes, Bulletin of Scientific Research, Vol. 3, pp. 51-58.
36. M. K. Panigrahi, (2022), Investigation of Structures of Sintered Fly Ash Materials: Resources of Industrial Wastes, Bulletin of Scientific Research, Vol. 4, pp.1-10.
37. M. Panigrahi, R. I. Ganguly, R. R. Dash, (2023), Development of pond ash (PA)-high carbon ferrochrome (HCFC) slag based geopolymer cementious materials, Chapter-5, ISBN-13:9781394166527, Publisher: Wiley.
38. Z. Cao, L. Xiang, E. Peng, and K. Li, (2018), Experimental Study on Electrical Resistivity of Cement-Stabilized Lead-Contaminated Soils, Geomaterials in Geotechnical Engineering, Vol. 2018, pp. xxx-xxx (https://doi.org/10.1155/2018/4628784).
39. F. Farhana, H. Kamarudin, A. Rahmat, Mohd M. Al Bakri Abdullah, and S. R. Shamsudin, (2015), The Electrical Resistivity of Geopolymer Paste by Using Wenner Four Probe Method, Key Engineering Materials, Vol. 660:pp. 28-33 (DOI: 10.4028/www.scientific.net/KEM.660.28).
40. Jingming Cai, Jinlong Pan, Xiaopeng Li, Jiawei Tan, and Jiabin Li, (2020), Electrical resistivity of fly ash and metakaolin based geopolymers, Construction and Building Materials, Vol. 234(2),pp. 117868-xxx (DOI: 10.1016/j.conbuildmat.2019.117868).
41. Walter J, Uthayakumar M, Balamurugan P, Mierzwiński D. (2021), The Variable Frequency Conductivity of Geopolymers during the Long Agieng Period. Materials (Basel). Vol. 14, pp. 5648-17 (doi: 10.3390/ma14195648).

8

High Resistant Exposed HCFC Slag Geopolymer

Muktikanta Panigrahi[1]*, Ratan Indu Ganguly[2] and Radha Raman Dash[3]

[1]*Department of Materials Science, Maharaja Sriram Chandra Bhanja Deo University, Keonjhar Campus, Odisha, India*
[2]*Department of Metallurgical Engineering, National Institute of Technology, Rourkela, Odisha, India*
[3]*CSIR-National Metallurgical Laboratory (NML), Jamshedpur, Jharkhand, India*

Abstract

Sustainable development of geopolymer as a cementitious material will benefit construction sectors. Geopolymer has been prepared from high-carbon ferrochrome (HCFC) slag, which is a waste material from the ferrochrome industry. Use of HCFC slag has advantages because of its low rate of greenhouse gas emissions.

In the conventional method of preparation, high-carbon ferrochrome (HCFC) slag geopolymer (GP) is synthesized by treating slag with alkali liquid solution. This enables binding non-reactive materials present in the slag. A cubical mold is used to cast green HCFC slag GP. Cubical GP product is cured at 70 °C for 24 h. Cured cubical GP sample is exposed under normal water up to 365 days for observing corrosion behavior of materials. It is observed that strength of geopolymer did not change much (before and after exposure).

The XRD analyses indicate that there is gradual transformation of crystalline phases to non-crystalline glassy phases. And the SEM images show the presence of a greater amount of glassy phase after curing, which is also corroborated with mechanical properties such as compressive strength. A TEM micrograph distinctly shows hexagonal shaped body with fused mass.

The TGA results are also discussed for both uncured and cured GP samples. DSC isotherms indicate oozing out of inbuilt water present in the prepared GP material, which is an indication of condensation polymerization reaction occurring during formation of geopolymer.

**Corresponding author*: muktikanta2@gmail.com

Muktikanta Panigrahi, Ratan Indu Ganguly and Radha Raman Dash (eds.) High Electrical Resistant Materials: Ferrochrome Slag Resource Ceramics, (177–196) © 2024 Scrivener Publishing LLC

Electrical resistance value (resistivity value) of HCFC slag GP (exposed and unexposed GP materials) is estimated by two-probe method.

Keywords: Industrial waste, HCFC slag, geopolymer, XRD, SEM, TEM, compressive strength, electrical resistivity, corrosion study

8.1 Introduction

In last few decades, Portland cement (PC) has become a common structural material used for construction of structures [1–4]. Portland cement production creates environmental issues, which is due to the release of carbon dioxide (CO_2) into the environment [5]. Manufacture of PC is an energy-intensive process [6, 7]. Therefore, research communities have utilized wastes such as fly ash (FA) [8], pond ash (PA) [9], slag [10], bottom ash (BA) [11], and quarry dust (QD) [12], etc., as resource materials for production of geopolymer [8–12]. Hence, researchers are trying to develop new-generation materials, i.e., GP to replace PC. One of these wastes is obtained from ferroalloy industries. Slag generated by these industries contain oxides such as magnesium oxide, iron oxide, chromium oxide (which exist in more than one oxidation state), and other oxides. Ferrochrome slag may contain elemental chromium. Major constituents of slag are CaO, SiO_2, Al_2O_3, MgO, and FeO. In the years 2008 and 1988, Davidovits forecasted GP to be a future constructional material, which would replace PC in constructional areas [13, 14]. Available reports indicate use of fly ash, pond ash, bottom ash, slag, etc., are suitable for production of geopolymer [8–12]. Geopolymers are oxides which form chain-like structure by reaction with activated alkali solution such as sodium hydroxide (NaOH) or potassium hydroxide (KOH) in combination with sodium silicate [14, 15]. Therefore, GP needs to be considered as an innovative constructional material for the 21st century [8–15]. Geopolymer structure has been described in many studies [16, 17]. Thus, the advantages of using GP materials lie in the fact that they are cheaper, non-hazardous, fire resistant and environmentally friendly [10].

Slag is waste obtained from the ferroalloy industries during the production of ferrochrome [10, 18, 19]. Slag is an oxide-bearing materials such as magnesium oxide, iron oxide and chromium oxide.

Therefore, the development of a low-cost GP material from locally available industrial waste has become an active field of research [8–15]. Currently, more attention is being paid to its development for use in insulation, encapsulation, multilayer ceramic chips, printed circuit boards, etc.

Reports are available where such materials have been developed for electronic applications [20–22]. Fly ash-based composite and biofiber-based composite have low dielectric constant and less dielectric loss at higher frequencies of applied field. Therefore, they are considered suitable for electronic applications. Composite is now prepared from coir dust (biowaste) and fly ash (industrial waste) adopting hand lay-up method [23–25].

Attempts are being made to make zeolite-based microporous materials. Zeolite mobilizes extra framework cations to induce ionic conduction [26].

Cui *et al.* [27] demonstrated the synthetization of geopolymer possessing high electrical conductivity of 1.5×10^{-6} S/cm at room temperature. Electrical conduction plays an important role for fast ionic conduction used in solid-state batteries and electrochemical sensors. Fabrication of geopolymer is an efficient method where the material possesses unique properties such as ionic conduction. This is because geopolymer contains water molecules and hydroxide, which greatly influence electrical conductivity and dielectric property at room temperature [28].

Kljajević *et al.* [29] synthesized geopolymers (GP) from metabentonite (MB), metadiatomite (MD) and metakaolinite (MK). Different test methods, such as XRD, FTIR, and SEM/EDX, were adopted to characterize synthesized GP. Geopolymers are quasi-solid electrolytes and possess high electrical conductivity. The highest conductivity is found for the sample obtained from geopolymer metakaolinite (GPMK) material which is 2.14×10^{3} Ω.cm (at 700 °C). Activation energies for electronic conduction have been calculated to be 0.33 eV in the temperature range of 500–700 °C. Geopolymer synthesized from metakaolin is recommended for use as an alternative material for solid oxide fuel cell (SOFC).

Farhana *et al.* [30] measured corrosion rate of geopolymer paste using Wenner four-probe methods. Rectangular shaped geopolymer sample was prepared from aluminosilicate materials, i.e., fly ash and an alkaline activator, for conduction of the test. They concluded from their work that corrosion rate of GP paste is negligible, i.e., very low for material even after exposing for a long time in corrosion medium.

Polder [31] assessed the corrosion rate of concrete material under corrosive environment. Usually, the corrosion process is initiated by chloride penetration and subsequent propagation of crack with time (corrosion rate) through the structure. Measurement of resistivity change of the structure will indicate corrosion rate. Low resistivity value is an indication of rapid chloride penetration and high corrosion rate. Once damage occurs, resistivity becomes relevant data for electrochemical maintenance methods. Resistivity of concrete can be measured non-destructively by placing electrodes on the concrete surface, applying a voltage and measuring the current. Several techniques can be used for measuring electrical resistivity. Practical aspects of these arrangements are discussed in detail. Procedures for calibration are proposed and guidelines for interpretation and reference values for various exposures and concrete compositions are given.

Song and Kwon [32] prepared reinforced concrete structures. Concrete structures have the potential to withstand different environmental conditions. However, failures in the structures still occur as a result of premature reinforcement corrosion. In order to maintain and repair bridges/buildings (for their safety) proper inspection is needed via monitoring techniques to assess reinforcement corrosion. An electrochemical method is used to assess corrosion. Electrochemical method is also applied for inspecting bridges, buildings and other civil engineering structures.

Glass *et al.* [33] protected steel in concrete through conductive coatings form using cathodic protection method. A property of such coating is evaluated as low sheet resistance (square resistance). Current density and anode connection geometry determines uniformity of current distribution over concrete surface. Sheet resistance is estimated by four-probe technique using (p/ln2)6(DV/I), which is non-destructive testing.

Corrosion tests and pH measurements have revealed that GP material has low leachability and excellent frictional resistance. Therefore, it can be used for construction of roads [34–37].

The present investigation is for preparing HCFC slag geopolymer from ferrochrome industry waste, i.e., slag. An attempt is made to characterize HCFC slag-based material through different techniques such as mechanical, structural, morphological, electrical analyses and corrosion behavior.

8.2 Experiment Details

8.2.1 Chemicals and Materials

For the current investigation, HCFC slag was procured from Shyam Ferro Alloys (Durgapur, West Bengal, India). Composition of HCFC slag is shown in Table 8.1. Other ingredients used were sodium hydroxide (NaOH) and sodium silicate (Na_2SiO_3), which were procured from Loba Chemie, India and Merck, India, respectively. Water-soluble plasticizer (Sika) was used as a plasticizer and was obtained from a reputed firm located in the southern part of India (Visakhapatnam market).

Table 8.1 Chemical composition of HCFC slag [10, 38].

Constituents of HCFC slag with percentages	SiO_2 = 28–31 MgO = 26.0–26.5 Al_2O_3 = 23.5–24.5 CaO = 7.5–8.5 Cr_2O_3 = 4.5–5.4 FeO = 2.5–3.0

Source: Shyam Ferro Alloys, Durgapur.

8.2.2 Geopolymer Synthesis

The steps for geopolymer preparation (from HCFC slag) are described below [9, 10].

Step 1: Raw material preparation
The HCFC slag is collected in the form of lumps. The lumps are crushed and ground in a ball mill. Ground HCFC slag is sieved through mesh (size 240). Sieved slags are dried in an oven at 120 °C for 2 h. Such slag is known as processed slag.

Alkaline activator solution is prepared using sodium silicate and sodium hydroxide in appropriate proportion. Sodium silicate solution is mixed with 8 M solution of sodium hydroxide to form alkaline activator solution. The solution is stirred continuously in a magnetic stirrer. Plasticizer (Sika) and extra water are mixed with the above prepared solution.

Step 2: Geopolymer preparation from raw materials
Processed HCFC slag is poured into a container. Then, alkali liquid solution is added slowly to the slag. It is mixed manually. The mixture is left for five minutes, during which time polymerization reaction occurs.

Step 3: Molding, casting and compaction
The mixture samples are poured into a cylindrical shaped mold (70 mm height and 55 mm diameter). Materials contained in the mold are rammed 15 times for their compaction. The whole mixture is allowed to set for 3 minutes before demolding samples. GP samples are removed from molds and are used for further investigations. Such GP samples are known as green samples. The entire process is illustrated in Figure 8.1.

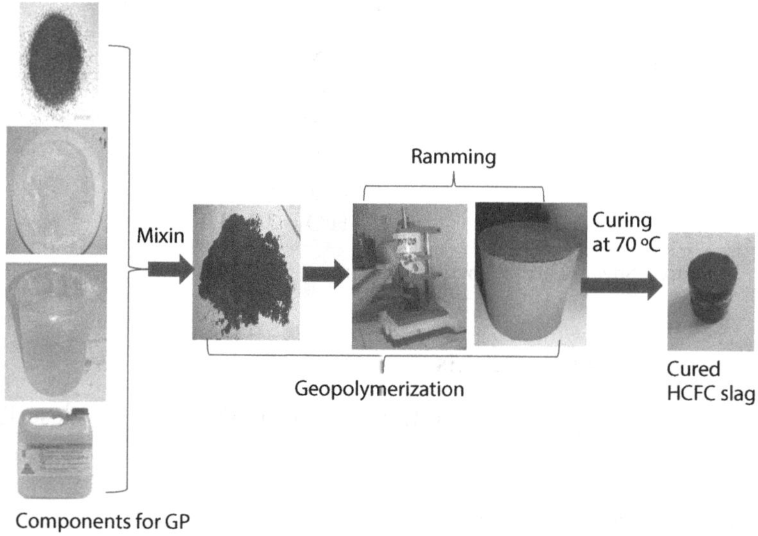

Figure 8.1 Schematic diagram of preparation process for HCFC slag-based GP product [10, 38].

Step 4: Curing
In the curing process, green samples are kept for heating in an oven at 70 °C for 24 h [9, 10]. Cured specimen(s) are used for different tests (such as XRD, FTIR TGA, DSC, etc.).

8.2.3 Test Methods

Various phases of HCFC slag-based materials are identified using a X-ray diffractometer. For this purpose, ferrochrome slag powder is dried at 110 °C for 24 h. It is further ground into the finer powder required for XRD analyses. Powder sample is placed on a quartz sample holder and is scanned through diffraction angle 2θ between 10° to 60° with a scanning speed of 5°/min. For this analysis, a Philips PW-1710 Advance wide-angle X-ray diffractometer coupled with a Philips PW-1729 X-ray generator is used.

The FTIR test method is used to analyze the presence of functional groups and formation of chemical linkage. A Thermo Nicolet Nexus 870 spectrophotometer is employed for this test. Instrument parameters are kept constant (50 scan at 4 cm^{-1} resolution, absorbance mode). For FTIR analyses, a 13 mm diameter, 0.3 mm thick pellet sample is prepared using KBr power and small amount of test sample. Before running the sample, a background spectrum is collected. Then, the pellet sample is put into a sample holder. Pellet is exposed to IR radiation.

Surface morphology of HCFC slag-based powder is analyzed by scanning electron microscope (Carl Zeiss Supra-40 model). Non-conducting samples need a thin layer of conducting coating. This is done by sputter coater. The elemental analyses are carried out with the help of EDX attached to the microscope.

A TEM microscope (model 5022/22, Tecnai G220 S-Twin, Czech Republic) is used to determine microscopic details at higher magnification and better resolution. Along with TEM studies, SAD patterns are examined for as-received HCFC slag geopolymer. Test specimens are transferred to carbon-coated Cu TEM grids.

In order to determine the stability of FeCr slag GP under corrosive atmosphere, a corrosion test is carried out. The GP products are initially weighed and then they are dipped into containers (5-liter capacity). Containers are filled with normal water. Samples are exposed for different lengths of time. After a regular interval, samples are removed in different batches. Samples are dried by exposing them in a heating oven at 120 °C. Then, the weights of dried samples are measured.

Electrical resistivity is measured by two-probe method. Exposed HCFC slag GP sample is used. Estimated resistivity is found using the following equation [39, 40]:

$$\rho = \left(\frac{R \times A}{l} \right) \tag{8.1}$$

Here, A is the area of cross-section $= \frac{\pi}{4} D^2$ (cm^2), R is the resistance of prepared samples (GΩ), l is thickness of the sample, and ρ is the resistivity of the samples in Ω.cm. For this measurement, Keithley source multimeter is used.

8.2.4 Results and Discussion

Three materials, namely HCFC slag (Figure 8.2a), unexposed cured HCFC slag GP (Figure 8.2b), and cured HCFC slag GP dipped in normal water for 365 days (Figure 8.2c), are shown in Figure 8.2. XRD analyses are made for each of these materials. Many identified phases are common. Some differences are observed between XRD patterns of as-received slag and HCFC slag GP. Relative intensity or peak heights of phases are reduced, indicating reduction in amount of crystalline phases for GP materials in comparison to those phases present in as-received material. There is a hump observed for GP sample close to 12° with considerable line-width. This is an indication for formation of amorphous phase in GP material, which is due to interaction between sodium silicate (SS) and sodium hydroxide (SH) with components present in HCFC slag.

New phases are formed (see Figure 8.2b). If XRD patterns of unexposed GP and exposed GP are compared, some differences are noticed in the pattern. Intensity or peak heights are different. The main phases of as-received slag identified are quartzite and mullite phase.

The SEM micrographs of raw HCFC slag, unexposed cured HCFC slag GP, and exposed cured HCFC slag GP in normal water are analyzed (Figure 8.3a, 8.3b and 8.3c). Figure 8.3a shows the SEM micrograph of raw HCFC slag. It consists of fused mass of oxides with some glassy phase spread over the matrix. Since the slag material is poured from high temperature to room temperature, faster cooling rate results in the formation of some

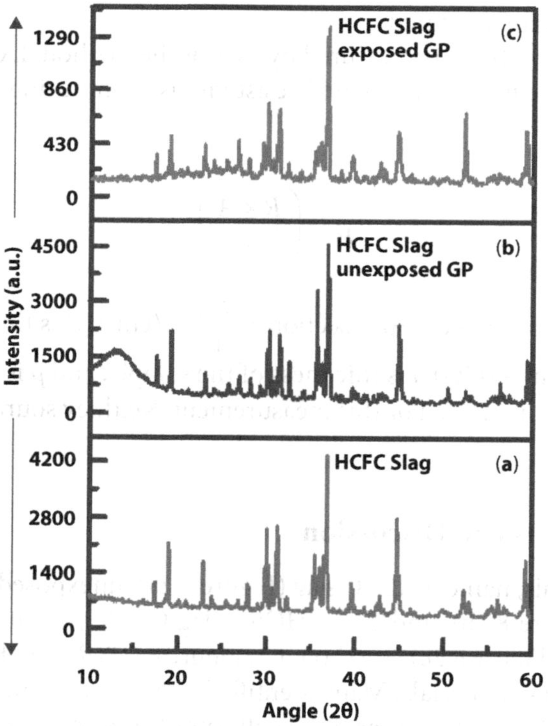

Figure 8.2 XRD patterns of raw HCFC slag (a), unexposed cured HCFC slag GP (b), and exposed cured HCFC slag GP in normal water (c).

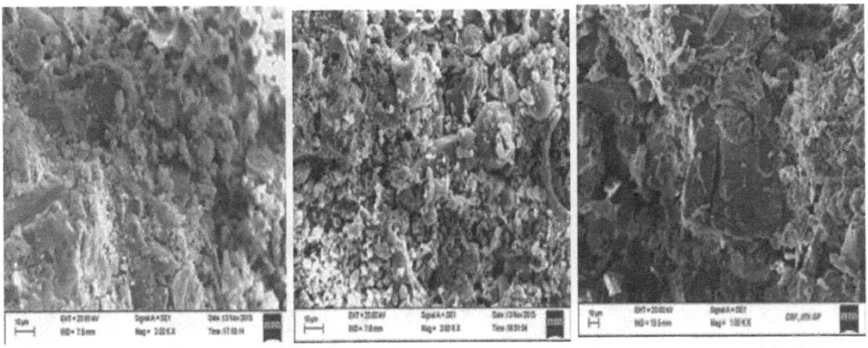

Figure 8.3 SEM image of raw HCFC slag (a), unexposed cured HCFC slag GP (b), and exposed cured HCFC slag GP in normal water (c).

amorphous phase. The micrograph in Figure 8.3b shows unexposed cured HCFC slag GP. This image shows fused crystalline phase. There is also glassy phase co-existing with crystalline phases. Figure 8.3c shows crystalline phase. The change in microstructures in GPs is due to crosslinking of alumino-silicate GP phase by reacting with sodium of sodium hydroxide.

Figure 8.4 shows the EDS spectra of raw HCFC slag, unexposed cured HCFC slag GP, and exposed cured HCFC slag GP in normal water. EDS analyses are carried out along with SEM studies. Elements such as O, Si, Al, Ca, Mg, and Cr are present in crystalline and glassy phases of all indicated samples (Figure 8.4a–c). However, the sodium element is absent in as-received HCFC slag. For GP samples (both exposed and unexposed), the sodium element is observed (Figure 8.4b and 8.4c), which is due to alkali treatment of raw material and formation of GP during polymerization process.

Figure 8.5a shows the TEM image of HCFC slag. In HCFC slag, hexagonal shaped crystals with some fused mass are observed in the matrix. Figure 8.5a also exhibits selected area diffraction (SAD) pattern of HCFC slag. Bright spots are concentrated around circumference of circles having different radius, which indicate the presence of crystalline phase in the as-received HCFC slag.

A TEM image of unexposed GP HCFC slag is shown in Figure 8.5b. Microstructure consists of a fused mass of presumably glassy phase with some crystalline phase. Selected area diffraction (SAD) pattern from the same area shows bright spots in different circumferences of circles. If the diffraction pattern of slag with GP is compared with electron diffraction of slag, it is observed that the number of bright spots is less, indicating the

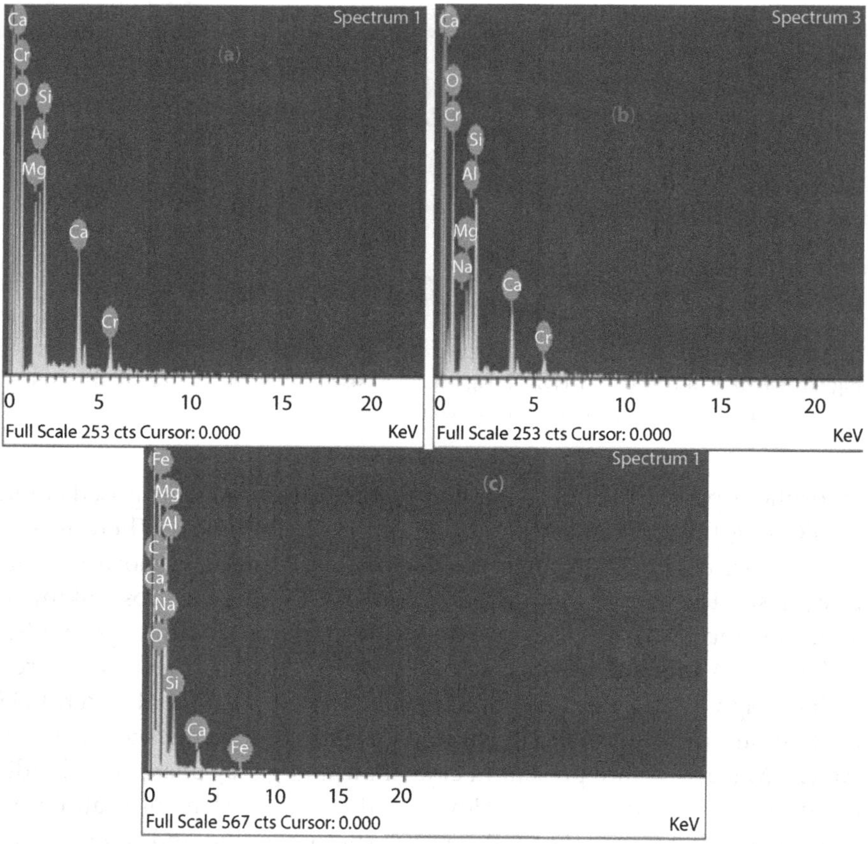

Figure 8.4 EDS spectra of raw HCFC slag (a), unexposed cured HCFC slag GP (b), and exposed cured HCFC slag GP in normal water (c).

amount of crystalline phases is less. This is because of the formation of GP from slag by alkali treatment.

After exposing GP material to normal water for 365 days, the microstructural features have not changed much (Figure 8.5b and 8.5c).

Electron diffraction (selected area diffraction) pattern of exposed cured HCFC slag GP in normal water is shown in Figure 8.5c'. Bright spots concentrated in fewer numbers have spread around the circumference of the circle. Hence, this exhibits crystalline phase of the samples.

Figure 8.6 shows FTIR spectra of HCFC slag, unexposed cured HCFC slag GP, and exposed cured HCFC slag GP in normal water. This indicates occurrence of different absorption bands at different wave numbers. Band at 3441 cm^{-1} wave number is in agreement with stretching vibration of O-H

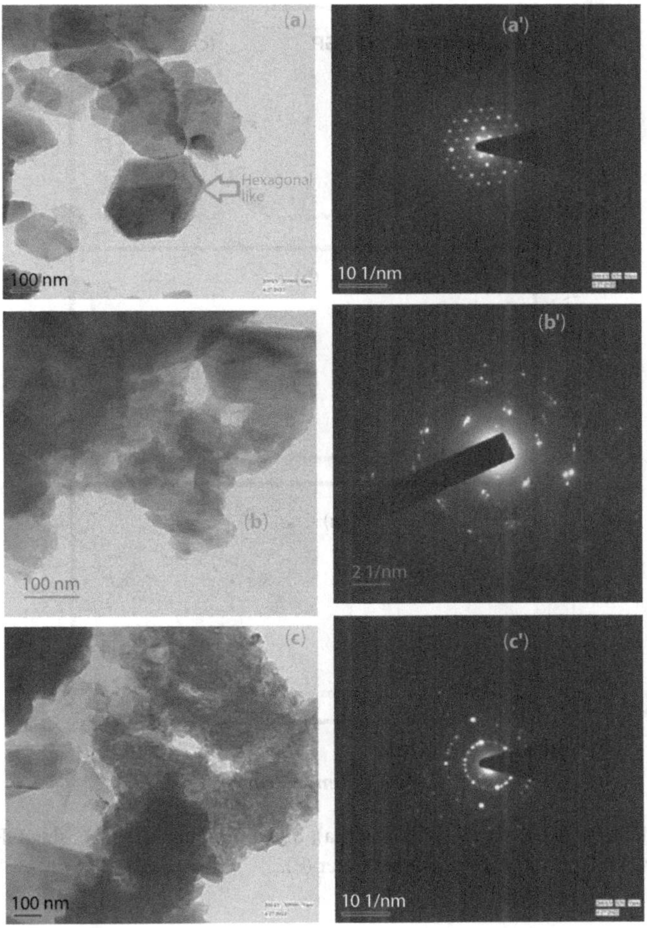

Figure 8.5 TEM images and SAD patterns of raw HCFC slag (a), unexposed HCFC slag GP (b), and exposed HCFC slag GP in normal water (c).

bonds. The H-O-H bending vibrations found at 1638 cm^{-1} are attributed to interlayer adsorbed H_2O molecule. Such hydroxyl-stretching band of water plays an important role in causing significant shifting of peak. The band at 887 cm^{-1} is attributed to Si-O band and signifies the occurrence of silicate groups. The $Al^{3+}O^{2-}$ band is indicated near 805 cm^{-1} wave number, and stretching band at 440 cm^{-1} signifies the existence of Fe-O band [41–46].

Figure 8.6b shows the FTIR spectrum of unexposed cured HCFC slag GP. A band is noticed at 3426 cm^{-1}, which is due to stretching vibrations of O-H bonds and H bending vibrations of interlayer adsorbed H_2O molecule. A C=O band is identified at 2270 cm^{-1}, which indicates the occurrence of

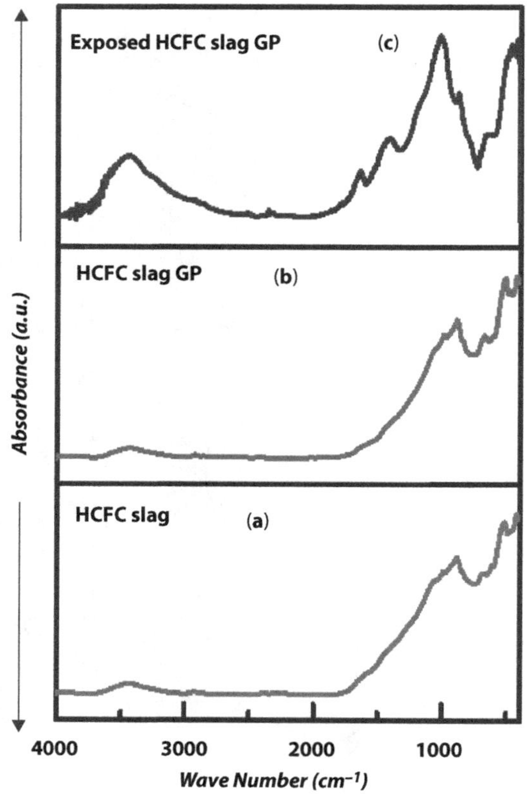

Figure 8.6 FTIR spectra of raw HCFC slag (a), unexposed cured HCFC slag GP (b), and exposed cured HCFC slag GP in normal water (c).

carbonate group. This is mainly attributable to the presence of chemisorbed CO_2 by unexposed cured HCFC slag GP. Si-O and O-Si-O bands are also found at 1071 cm^{-1} and 495 cm^{-1}, which specify silicate groups. The presence of Al^{3+}-O^{2-} band and Fe-O are also observed near 858 cm^{-1} and at 409 cm^{-1}, respectively [41–46].

The FTIR spectrum of exposed cured HCFC slag GP in normal water is shown in Figure 8.6c. In this figure, prominent broad band is observed at 3426 cm^{-1}, whcih is attributed to O-H bond. Band at 2270 cm^{-1} is identified to be C=O group, which indicates the presence of carbonate. Bands at 1071 cm^{-1} and 495 cm^{-1} have established the occurrence of Si-O and O-Si-O groups. These bands indicate the presence of silicate groups. Al^{3+}-O^{2-} and Fe-O bands are found at 858 cm^{-1} and at 409 cm^{-1}, respectively [41–48].

The bar diagram in Figure 8.7 depicts the weight in grams vs. number of days of unexposed and exposed (with different lengths of exposing time)

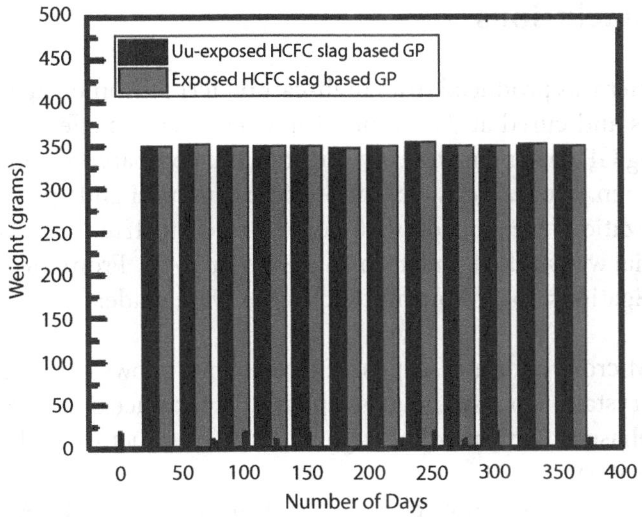

Figure 8.7 Corrosion results of HCFC slag GP cubical sample in normal water.

cured cubical HCFC slag GP sample(s) in normal water. It is found that there is no change in weight of exposed materials in comparison to unexposed materials. Therefore, it is concluded that GP materials are resistant to normal water.

Resistivity values of exposed HCFC slag GP (normal water) is determined by two-probe method (Table 8.2). Resistivity value (resistance value) is found to be 49.61×10⁹ Ohm.cm (3.51 GΩ). The resistivity value is due to formation of mullite phases, magnesium aluminate phase, and oxides of chromium-based complex phase [49].

Table 8.2 Resistivity value of unexposed and exposed HCFC slag GP (normal water).

S.no.	Sample ID	Resistivity measurement temperature (°C)	Resistance value (Gigaohm, GΩ)	Resistivity value (Ohm.cm)
1	Exposed HCFC slag GP (Normal water)	~30 °C	3.51	49.61×10^9

8.3 Conclusions

Geopolymer was produced from high-carbon ferrochrome slag using alkaline actors and cured at 70 °C for 24 h. Characteristic effects of exposed HCFC slag GP specimen were investigated and compared with HCFC slag GP specimen. Mechanical, electrical microstructural and spectroscopical characterizations were made with sophisticated instruments. Stability of GP material was studied under normal water (pH~7). From the findings of the investigations, the following conclusions were made:

- Microstructure of exposed geopolymer shows presence of crystalline phase and glassy phase. Occurrence of crystalline phase is confirmed through TEM microstructure and SAD pattern.
- EDS analyses show presence of O, Si, Al, Cr, Ca, Mg, Al, Na, etc., elements in exposed and unexposed HCFC slag GP materials.
- Compressive strength of exposed cured (at 70 °C in an oven for 24 h) HCFC slag GP sample is determined to be 16 MPa.
- FTIR analyses have revealed new bonds of exposed and unexposed HCFC slag GP materials if compared with as-received slag from which GP is prepared.
- TGA curves have revealed the stability of GP material in the temperature range of 50–800 °C. The DSC isotherm of exposed sample signifies condensation polymerization process occurring in the geopolymer.
- Resistivity value of exposed HCFC slag GP is determined to be 49.61×10^9 (Ohm,cm).
- Corrosion results indicate no reduction in weight of the specimens (before and after exposure).

Acknowledgments

The authors convey their sincere thanks to the Ministry of Mines, Government of India, for providing the grant to carry out the work [Grant number= F.No.:14/54/20214-Met.-IV; dated: 29.12.2014]. They also convey their sincere thanks to GIET, University Gunupur, Rayagada, Odisha, India for providing lab facilities to do the research work. Rhey would also like to thank CRF IIT Kharagpur for providing testing facilities.

The authors would like to thank Prof. Munesh Chandra Adhikary, PG Council Chairman, Fokir Mohan University, for his invaluable guidance, advice, and constant inspiration throughout the entire process. They would also like to thank Mr. Mukteswar Mohapatra of Fokir Mohan University for his support.

References

1. J. J. Biernacki, J. W. Bullard, G. Sant, N. Banthia, K. Brown, F. P. Glasser, S. Jones, T. Ley, R. Livingston, L. Nicoleau, J. Olek, F. Sanchez, R. Shahsavari, P. E. Stutzman, K. Sobolev, T. Prater, (2017), Cements in the 21st Century: Challenges, Perspectives, and Opportunities, Journal of American Ceramic Society, Vol. 100, pp. 2746-2773. (doi: 10.1111/jace.14948).
2. S. Shahi, E. Fakhri, H. Yavari, S. M. Dizaj, S. Salatin, and K. Khezri, (2022), Portland Cement: An Overview as a Root Repair Material, BioMed Research International, Vol. 2022, pp. 1-13 (https://doi.org/10.1155/2022/3314912).
3. X. Song, X. Song, H. Liu, H. Huang, K. G. Anvarovna, N. A. D. Ugli, Y. Huang, J. Hu, J. Wei and Q. Yu , (2022), Cement-Based Repair Materials and the Interface with Concrete Substrates: Characterization, Evaluation and Improvement, Polymers, Vol. 14, pp. 1485-xxx (https://doi.org/10.3390/polym14071485).
4. S. Fei, H. Huang, G. Kasimova, A. Nurmirzayev, Y. Huang, J. Hu, J. X. Wei, and Q. Yu, (2023), Cement-Based Repair Materials with Concrete Substrates. Encyclopedia. Available online: https://encyclopedia.pub/entry/21563 (accessed on 27 March 2023). (10.3390/polym14071485)
5. O. E. Ige, O. A. Olanrewaju, K. J. Duffy and O. C. Collins, (2022), Environmental Impact Analysis of Portland Cement (CEM1) Using the Midpoint Method, Energies, Vol. 15, pp. 2708-xxx (https://doi.org/10.3390/en15072708).
6. E. Worrell, L. Price, N. Martin, C. Hendriks, and L. O. Meida, (2001), Carbon Dioxide Emissions from the Global Cement Industry, Annual Review of Energy and the Environment, Vol. 26, pp. 303–29.
7. https://www.epa.gov/stationary-sources-air-pollution/portland-cement-manufacturing-industry-national-emission-standards.
8. N. B. Singh, (2018), Fly Ash-Based Geopolymer Binder: A Future Construction Material, Minerals, Vol. 8, pp. 299-xxx (https://doi.org/10.3390/min8070299).
9. M. Panigrahi, R. I. Ganguly, R. R. Dash, (2018), Development of Novel Constructional Material from Industrial Solid Waste as Geopolymer for Future Engineers, IOP Conference Series: Materials Science and Engineering, Vol. 410, pp.1-12

10. M. Panigrahi, R. I. Ganguly, R. R. Dash, (2018), An Advanced Cured High Carbon Ferrochrome Slag (HCFCS) Geopolymer (GP): A Constructional Materials, IOP Conference Series: Materials Science and Engineering, Vol. 410, pp. 1-15.
11. L. J. J. Nieves, F. Elyseu, S. Goulart, M. de S. Pereira, E. Z. Valvassori, and A. M. Bernardin, (2021), Use of fly and bottom ashes from a thermoelectrical plant in the synthesis of geopolymers: Evaluation of reaction efficiency, Energy Geoscience, Vol. 2, pp. 167-173 (https://doi.org/10.1016/j.engeos.2020.09.004).
12. E. Cohen, A. Peled, and G. Bar-Nes, (2019), Dolomite-based Quarry-Dust as a Substitute for Fly-Ash Geopolymers and Cement Pastes, Journal of Cleaner Production, Vol. 235, pp. 910-919
13. J. Davidovits, Geopolymer Chemistry and Applications. Edition: 2nd (2008), 3rd (2011), 4th (2015), 5th (2020), Publisher: Institute Geopolymer, Geopolymer Institute, Saint-Quentin, France, Editor: Joseph Davidovits, ISBN: 9782954453118 (5th ed.)
14. J. Davidovits, (1988), Soft Mineralogy and Geopolymers. Proc. 1st Int. Conf. on Geopolymers, Compiegne, France, Vol. 1, pp. 19-23.
15. A.M.M. Al Bakri, H. Kamarudin, M. Bnhussain, I. K. Nizar, and W.I.W. Mastura, (2011), Mechanism and Chemical Reaction of Fly Ash Geopolymer Cement- A Review, Asian Journal of Scientific Research, Vol. xxx,
16. I. Giannopoulou and D. Panias, (2007), Structure, Design and Applications of Geopolymeric Materials Conference: 3rd International Conference on Deformation Processing and Structure of Materials, Vol. xxx, pp. 5-15.
17. J. G. S. van Jaarsveld and J. S. J. van Deventer, (1999), Effect of the Alkali Metal Activator on the Properties of Fly Ash-Based Geopolymers, Industrial & Engineering Chemistry Research, Vol. 38, pp. 3932–3941 (https://doi.org/10.1021/ie980804b).
18. P. H. Kumar, A. Srivastava, V. Kumar, M. R. Majhi, V. K. Singh, (2014), Implementation of Industrial Waste Ferrochrome Slag in Conventional and Low Cement Castables: Effect of Microsilica Addition, Journal of Asian Ceramic Societies, Vol. 2, pp. 169-175 (https://doi.org/10.1016/j.jascer.2014.03.004).
19. B. A. V. R.Kumar, L. Keshav, P. A. Sivanantham G. G. V. Arokiaraj, D. R. Zunaithur Rahman, P. M. Kumar, and D. Somashekar (2022), Comprehensive Characterization of Ferrochrome Slag and Ferrochrome Ash as Sustainable Materials in Construction, Vol. 2022, pp.1-12 (https://doi.org/10.1155/2022/8571055).
20. N. A. Mohd Mortar, M. M. Al Bakri Abdullah, R. Abdul Razak, S. Z. Abd Rahim, I. H. Aziz, M. Nabiałek, R. P. Jaya, A. Semenescu, R. Mohamed, and Mohd F. Ghazali, (2022), Geopolymer Ceramic Application: A Review on Mix Design, Properties and Reinforcement Enhancement, Materials, Vol. 15, pp. 7567 (https://doi.org/10.3390/ma15217567).

21. A. Saeed, H. Mohammed Najm, A. Hassan, Mohanad M. S. Sabri, S. Qaidi, N. S. Mashaan, and K. Ansari, (2022), Properties and Applications of Geopolymer Composites: A Review Study of Mechanical and Microstructural Properties, Materials, Vol. 15, pp. 8250-xxx (https://doi.org/10.3390/ma15228250).
22. F. Somidin, N. F. N. Ilias, N. S. Mohamad Zaimi, Mohd I. I. Ramli, and N. R. Abdul Razak, (2022), Influence of Fly Ash Geopolymer Ceramic Powder Addition on Sn-3.0Ag-0.5Cu Solder Joints, Journal of Physics: Conference Series, Vol. 2169, pp. 012032-xxx (doi:10.1088/1742-6596/2169/1/012032).
23. G.-A. M Nasr and R. M. Ahmed, (2020), Characterization and Dielectric Properties of Fly Ash / Polystyrene / Low Density Polyethylene Composites, IOP Conference Series Materials Science and Engineering Vol. 956, pp. 012002 (DOI: 10.1088/1757-899X/956/1/012002).
24. J. P. Dhal, and S. C. Mishra, (2013), Investigation of Dielectric Properties of a Novel Hybrid Polymer Composite Using Industrial and Bio Waste, Journal of Polymer & Composites, Vol. 1, pp. 22–27.
25. https://nhsjs.com/2012/development-of-a-novel-hybrid-polymer-composite-with-coir-dust-and-fly-ash-reinforcements-and-investigation-of-its-dielectric-nature/
26. M. Khosravi, V. Murthy, and I. D. R. Mackinnon, (2019) The Exchange Mechanism of Alkaline and Alkaline-Earth Ions in Zeolite-N, Molecules, Vol. 24, pp. 3652-xxx. (doi: 10.3390/molecules24203652).
27. X.-M. Cui, G.-J. Zheng, Y.-C. Han, F. Su and J. Zhou, (2008), A study on Electrical Conductivity of Chemosynthetic Al_2O_3–$2SiO_2$ Geopolymer Materials, Journal of Power Sources 184(2):652-656 (DOI: 10.1016/j.jpowsour.2008.03.021).
28. N. Chuewangkam, T. Nachaithong, N. Chanlek, P. Thongbai, and S. Pinitsoontorn, (2022), Mechanical and Dielectric Properties of Fly Ash Geopolymer/Sugarcane Bagasse Ash Composites, Polymers (Basel), Vol. 14, pp. 1140-xxx (doi: 10.3390/polym14061140).
29. L. M. Kljajević, Z. Melichová, Z. Melichová, M. Stojmenović, M. Stojmenović, S. Nenadović, S. Nenadović, (2019), Structural and Electrical Properties of Geopolymer Materials based on Different Precursors (Kaolin, Bentonite and Diatomite), Macedonian Journal of Chemistry and Chemical Engineering, Vol. 38, pp. 283-292 (DOI: 10.20450/mjcce.2019.1678).
30. F. Farhana, H. Kamarudin, A. Rahmat, M. M. Al Bakri Abdullah, S. R. Shamsudin, M. T. Selimin, and A. V. Sandu, (2015), The Electrical Resistivity of Geopolymer Paste by Using Wenner Four Probe Method, Key Engineering Materials, Vol. 660, pp. 28-33 (DOI: 10.4028/www.scientific.net/KEM.660.28)
31. R. B Polder, (2001), Test Methods for on-Site Measurement of Resistivity of Concrete-A RILEM TC-154 Technical Recommendation, Construction and Building Materials, Vol. 15, pp. 125-131 (https://doi.org/10.1016/S0950-0618(00)00061-1).

32. H.-W. Song, and S.-J. Kwon, (2007), Permeability Characteristics of Carbonated Concrete Considering Capillary Pore Structure, Cement and Concrete Research, Vol. 37, pp. 909-915 (https://doi.org/10.1016/j.cemconres.2007.03.011).
33. G. K. Glass, A. M. Hassanein, and N. Buenfeld, (2000), CP Criteria for Reinforced Concrete in Marine Exposure Zones, Journal of Materials in Civil Engineering, Vol. 12, pp. xxx-xxx (DOI: 10.1061/(ASCE)0899-1561(2000)12:2(164).
34. A. A. S. Tigue, Roy A. J. Malenab, J. R. Dungca, D. E. C. Yu, and M. Angelo B. Promentilla, (2018), Chemical Stability and Leaching Behavior of One-Part Geopolymer from Soil and Coal Fly Ash Mixtures, Minerals, Vol. 8, pp. 411-xxx (https://doi.org/10.3390/min8090411).
35. W. Yang, P. Zhu, H. Liu, X. Wang, W. Ge, and M. Hua, (2021), Resistance to Sulfuric Acid Corrosion of Geopolymer Concrete Based on Different Binding Materials and Alkali Concentrations, Materials, Vol. 14, pp. 7109-xxx (https://doi.org/10.3390/ma14237109).
36. T. Lingyu, H. Dongpo, Z. Jianing, W. Hongguang, (2021), Durability of geopolymers and geopolymer concretes: A review, Reviews on Advanced Materials Science, Vol. 60, pp. 1–14 (DOI:10.1515/rams-2021-0002).
37. M. F. Ghazy, M. H. Taman, and S. S. ELatfawy, (2022), Properties and Durability of Geopolymer Based Materials and It's Utilizing as a Repair and Strengthening Materials: A Review, International Research Journal of Engineering and Technology (IRJET), Vol. 09, pp. 1272- 1290.
38. M. Panigrahi, S. Mohanty, R. I. Ganguly, R. R. Dash, (2021), Synthesis and Characterizations of High Carbon Ferrochrome (HCFC) Slag based Geopolymer, IntechOpen Publisher, UK. ISBN: 978-1-83968-989-5, Vol. xxx, pp. 1-15 (DOI: http://dx.doi.org/10.5772/intechopen.97140).
39. M. K. Panigrahi, (2021), Investigation of Structural, Morphological, Resistivity of Novel Electrical Insulator: Industrial Wastes, Bulletin of Scientific Research, Vol. 3, pp. 51-58.
40. M. K. Panigrahi, (2022), Investigation of Structures of Sintered Fly Ash Materials: Resources of Industrial Wastes, Bulletin of Scientific Research, Vol. 4, pp.1-10.
41. B.J. Saikia, G.K Rao, and P. Sarathy, (2010), Fourier Transform Infrared Spectroscopic Characterization of Kaolinite from Assam and Meghalaya, North Eastern India, International Journal of Modern Physics., Vol. 1, pp. 206.
42. R. P. dos Santos, J. Martins, C. Gadelha, B. Cavada, A.V. Albertini, F. Arruda, M. Vasconcelos, E. Teixeira, F. Alves, J.L. Filho, and V. Freire, (2014), Coal Fly Ash Ceramics: Preparation, Characterization, and Use in the Hydrolysis of Sucrose. Scientific World Journal, Vol. 2014, pp. 01.
43. A. Jose, M. R. Nivitha, M. Krishnan, and R. G. Robinson, (2020), Characterization of Cement Stabilized Pond Ash using FTIR Spectroscopy, Construction and Building Materials, Vol. 263, pp. 120136-xxx (DOI: 10.1016/j.conbuildmat.2020.120136).
44. Y. Liu, F. Zeng, B. Sun, P. Jia, and I. T. Graham, (2019), Structural Characterizations of Aluminosilicates in Two Types of Fly Ash Samples from

Shanxi Province, North China, Minerals, Vol. 9, pp. 358-16 (doi:10.3390/min9060358).
45. P. Rożek, M. Król, and W. Mozgawa, (2018), Spectroscopic Studies of Fly Ash-Based Geopolymers, Spectrochimica Acta - Part A Molecular and Biomolar Spectroscopy, Vol. 198, pp. 283-289 (https://doi.org/10.1016/j.saa.2018.03.034).
46. S. Kumar, F. Kristály, and G. Mucsi, (2015), Geopolymerization Behaviour of Size Fractioned Fly Ash, Advanced Powder Technology, Vol. 26, pp. 24-30 (https://doi.org/10.1016/j.apt.2014.09.001).
47. Z. Kledyński, A. Machowska, B. Pacewska and I. Wilińska, (2017), Investigation of Hydration Products of Fly Ash-Slag Pastes, Journal of Thermal Analysis and Calorimetry, Vol. 130, pp. 351-363 (https://doi.org/10.1007/s10973-017-6233-4).
48. D. Bondar and R. Vinai, (2022), Chemical and Microstructural Properties of Fly Ash and Fly Ash/Slag Activated by Waste Glass-Derived Sodium Silicate, Crystals, Vol. 12, pp. 913-12 (https:// doi.org/10.3390/cryst12070913).
49. S. Montayev, B. Shakeshev, and S. Zharylgapov, (2020), Development of a Technology for Producing Ceramic Refractory Material in a Composition of Montmorillonite Clays (Bentonite-Like) and Ferrochrome Production Wastes, MATEC Web of Conferences, Vol. 315, pp. 07007 (DOI: 10.1051/matecconf/202031507007).

9

High Resistant HCFC Slag/ Fly Ash Hybrid Ceramics

M. K. Panigrahi[1]*, R. I. Ganguly[2] and R. R. Dash[3]

[1]Department of Materials Science, Maharaja Sriram Chandra Bhanja Deo University, Keonjhar Campus, Odisha, India
[2]Department of Metallurgical Engineering, National Institute of Technology, Rourkela, Odisha, India
[3]CSIR-National Metallurgical Laboratory, Jamshedpur, Jharkhand, India

Abstract

Fly ash is a by-product of thermal power plants, which is highly accountable for environmental pollution. Hence, an attempt has been made to develop a high-resistance material from hybrid ceramic composite consisting of FA, HCFC slag, CC, PY, k-FD, and BC. The process involves sintering of the mixture at high temperature (i.e., 1200 °C).

Electrical resistance value of the sintered material is measured by two-probe technique. The highest resistance value (resistivity value) is found to be 36.93 GΩ (521.99×10^9 Ω.cm). Vickers hardness value of sintered hybrid composite is measured to be 50.052.

Sample has been characterized by X-ray diffraction (XRD), Fourier-transform infrared (FTIR) spectroscopy, optical microscopy, scanning electron microscopy (SEM), and transmission electron microscopy (TEM) to determine qualitatively structure. The above studies have revealed the presence of different chemical constituents, chemical groups and surface topography. All these studies have enabled us to understand the mechanism of formation of products.

Phases like quartz, mullite, forsterite, spinel, chromium-bearing phase, etc., are identified. Among the phases, the mullite phase has contributed the most towards the enhancement of insulating properties of sintered hybrid composite. An optical micrograph of hybrid sample shows the presence of particles marked by white patches spread over the texture. The SEM microstructure shows crystalline and glassy phases (fused mass), which are formed during sintering at high temperature.

**Corresponding author*: muktikanta2@gmail.com

Muktikanta Panigrahi, Ratan Indu Ganguly and Radha Raman Dash (eds.) High Electrical Resistant Materials: Ferrochrome Slag Resource Ceramics, (197–216) © 2024 Scrivener Publishing LLC

The TEM image displays fused mass crystals spread over the texture. In the SAD pattern, bright spots of sintered composite product indicate the presence of crystalline phase in the microstructure. Along with SEM, EDS analysis is also done to assess occurrence of elements in the sintered material. Different chemical groups of samples are made by FTIR analyses.

***Keywords*:** Fly ash, HCFC slag, kaolin, pyrophyllite, k-feldspar, ball clays, mullite phases, insulator

9.1 Introduction

Currently, coal plays a vital role in generation of thermal power. This has been the practice over the last few decades. As such, the carbon content of coal is gradually decreasing, thereby generating a large quantity of ash. Ash is classified into two groups, i.e., fly ash and bottom ash. Disposal of these materials has become a problem for thermal power plants. Additionally, environmental pollution has arrived at a threatening point. Governments are concerned about the present situation and are providing incentives to scientists in order to discover fruitful ways to utilize these materials. Alternatively, other means of power generation are being actively considered in order to restrict the use of coal for power generation.

Since global power requirements are still largely met by coal, a large amount of ashes continue to accumulate [1–8]. Therefore, fly ash has recently begun being reused in different manufacturing processes such as construction of roads, manufacture of bricks, production of ceramics, etc. Bottom ash and pond ash are yet to be used fruitfully. Fly ash is one of the by-products collected by electrostatic precipitators, mechanical filters, or simply ejected into the atmosphere [9]. It is composed of hollow spherical particles known as cenospheres. Hollow spherical particles are of low density and consist of glassy aluminum-silicate matrix, quartz, mullite, calcite, and magnetite [10]. Usually, 70% of fly ash is generated by thermal power plants during combustion of coal [11]. Because they pollute Earth's soil, water, and air, the presence of these wastes in the environment is detrimental to living beings [12].

Ceramics are used for different applications, such as refractories [13], cutting tools [14], thermal insulators [15], etc., which are effectively used by industries.

Preparation of ceramics involves application of load (e.g., pressure) followed by firing/sintering at higher temperature. Since raw materials contain alumina and silica, sintering between temperature 1200–1300 °C helps in the formation of crystallized phase, i.e., mullite phase. Mullite phase helps to enhance mechanical and electrical resistivity of ceramics [16].

For construction of electrical air lines, electrical insulators made up of ceramic materials are widely used [17]. Advantages of ceramic materials lie in the fact that they possess good mechanical strength, high dielectric strength, and good corrosion resistance [18, 19]. Use of such material will be cost-effective and can meet the demands of the electrical engineering sector. Researchers are working to develop good electrical insulators from waste products, i.e., fly ash, slag, bottom ash, etc. Researchers are being encouraged to develop cheaper insulators with adequate properties using waste material [20]. This will also help to reduce environmental pollution and protect land from being dumped with such waste. Insulators have great importance in day-to-day life. They make life easier and are harmless and shock-free [21–23].

A few properties have been observed with sintered ceramics. With proper application of pressure, temperature, and particle size, it is possible to make dense ceramic products approaching zero water absorption and open porosity property [24]. Such ceramic product has high hardness, abrasion resistance and flexural strength. Usually, raw materials required for ceramic preparation are kaolin, feldspar quartz, pure quartz sand, and ball clays [25, 26]. These raw materials have good mineralogical composition. By mixing them with wastes, such as fly ash, pond ash, bottom ash, slags, etc., it will be possible to prepare cost-effective ceramics [27, 28].

The objective of the present work is to prepare sintered FA/CC/PY/k-FD/BC hybrid materials. Components of the hybrid composite material are mixed in requisite proportion. They are sintered/fired at 1200 °C. The sintered materials are analyzed by SEM and EDS. Different phases present in sintered material are identified using X-ray diffraction technique. FTIR analyses are done to identify chemical groups in samples. Room-temperature electrical resistivity value is also estimated.

9.2 Experiment Details

9.2.1 Materials and Chemicals

Fly ash was procured from NALCO (a Navratna CPSE under Ministry of Mines), Damanjodi, Odisha, India; whereas HCFC slag was procured from Shyam Ferro Alloys, Durgapur, West Bengal, India. Kaolin, dextrin $(C_6H_{10}O_5)_{n.}xH_2O$, k-feldspar, pyrophylite, and ball clays were procured from Merck, India. A few drops of water (i.e., 6%) were required during the preparation of green paste. Compositions of raw samples, such as processed fly ash, processed HCFC slag, kaolin, k-feldspar, pyrophyllite, and ball clays, are showed in Table 9.1. Optical images of raw samples (i.e., fly

Table 9.1 Constituents with percentage (%) of as-processed fly ash, processed HCFC slag, kaolin, pyrophyllite, k-feldspar, and ball clays.

S. ID	Compositions (%)												
	Fe_2O_3	Al_2O_3	SiO_2	P_2O_5	Cr_2O_3	ZnO	MgO	C	CaO	TiO_2	MnO	LOI	Ref
FA	3.85	28.3	62.8	0.32	0.04	0.027	0.049	1.15	0.7	1.84	0.03	0.5	[29]
HCFC Slag	2.5-3.0	24.5-23.5	28-31	---	4.5-5.4	---	26.0-26.5	---	7.5-8.5	---	---	---	[30]
CC	---	37.66	44.80	---	---	---	T	---	0.50	0.60	---	14.33	[31]
PY	---	27.81	65.95	---	---	---	0.64	---	0.022	0.12	---	5.92	[32]
FD	---	18.18	64.57	---	0.22	---	---	---	---	0.01	----	---	[33]
BC	0.90	27.0	58.0	---	---	0.2	<0.1	---	<0.1	1.5	---	---	[34]

Note: S. ID = Sample ID, FA = Fly Ash, HCFC slag = High-carbon ferrochrome slag, CC = Kaolin, PY = pyrophyllite, k-FD = k-Feldspar (Potassium), BC = Ball clays.

Figure 9.1 Optical images of fly ash (a), kaolin (b) and dextrin (c); pyrophyllite (d), k-feldspar (e), ball clay (f), and HCFC slag (g); SEM images of HCFC slag (h) and fly ash (i).

ash, HCFC slag, kaolin, dextrin, k-feldspar, pyrophyllite, and ball clays) are shown in Figure 9.1.

9.2.2 FA/HCFC Slag/CC/PY/k-FD/BC Sintered Hybrid Materials Reparation

Preparation of hybrid ceramics from FA/HCFC slag/kaolin/pyrophyllite/k-feldspar/ball clay is achievable through solid-state route followed by sintering process [31]. The steps used in the process are as follows:

Step 1: Raw materials preparation process
Collected fly ash is ground in a ball mill for 5 h to reduce its size. It is then screened through 240 meshes. Screened fly ash is dried in a heating oven

at 120 °C (2 h) to remove moisture. Dried fly ash is known as processed fly ash. A similar route is followed for processed HCFC slag.

Step 2: Green pellet preparation process
Processed fly ash (60%) is placed in a mortar with pestle as main component. Then HCFC slag, kaolin, pyrophyllite, feldspar (potassium), and ball clays amounting to 40% are mixed with 60% fly ash. Compositions of mixture are shown in Table 9.2. Minor components are added to fly ash one after the other in a regular interval of half an hour during the process of mixing and grinding. Dextrin (i.e., 0.05%) is added to the mixture, which helps to bind the mixture. A few drops of water are used to make the mixture pasty. Such paste samples are known as green samples. Samples are now ready for pellet.

Pellets are made by a pelletizer where pressure is applied in a universal testing machine (UTM). Before beginning the process, different parts of the pelletizer are properly cleaned, washed, and dried. A calculated amount of the above-prepared paste is placed in the pelletizer. Pelletizer with paste is positioned in the UTM machine. Load (10 MPa) is applied to the pelletizer during compaction. It is left for five minutes for proper compaction. Then, pelletizer is removed from the UTM and pellet sample is collected from the pelletizer. With the help of slide calipers, thickness and diameter of pellet(s) are estimated. Thickness is found to be 1.1 mm and diameter is found to be 10 mm. Such pellet is known as green pellet. It is then ready for use in the next step of the process.

Step 3: Sintering process
Green pellet is bisquetted in a muffle furnace at 900 °C for 2 h. Thereafter, temperature of furnace is increased to 1200 °C where bisquetted sample is sintered for 1 h. Samples are allowed to cool slowly to room temperature

Table 9.2 Compositions (%) of fly ash/HCFC slag/CC/PY/k-FD/BC hybrid materials preparation.

S.no.	**Individual composition name**					
	Fly ash	**Kaolin (CC)**	**Pyrophyllite (PY)**	**k-feldspar (k-FD)**	**Ball clays (BC)**	**HCFC slag (%)**
1	60	10	5	10	10	5
2	60	10	8	10	5	7
3	60	10	5	10	5	10

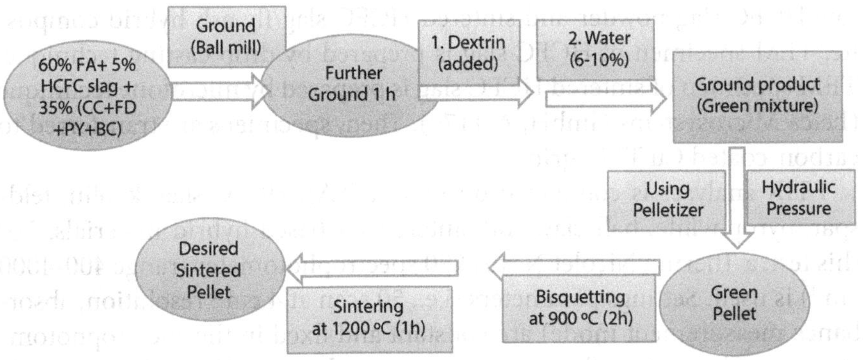

Figure 9.2 Flow chart for preparation of batch-1 [Fly ash (60%) + HCFC slag (5%) + 35% others (CC+k-FD+PY+BC) composite sintered at 1200 °C].

inside the furnace. Then, pellet is removed from the furnace. A dimension (i.e., thickness and diameter) of sintered pellet is measured with a slide caliper. Thickness and diameter are found to be 1.09 mm and 10 mm, respectively. The entire process is schematically shown in Figure 9.2.

Sintered pellet(s) is ready for different tests. Samples of different compositions are prepared and heat-treated in a similar manner as described earlier.

9.2.3 Test Methods

X-ray diffraction (XRD) analyses are carried out with a Philips PW-1710 Advance wide-angle X-ray diffractometer coupled with a Philips PW-1729 X-ray generator to identify the different phases. CuKα radiation having wavelength (λ) of 0.154 nm is used for diffraction studies. The XRD machine functions at 40 kV and 20 mA. Prepared pellet sample is put on a sample holder (made of quartz). Sample is scanned from 10° to 70° with scanning speed 5°/min.

Optical images of sintered HCFC slag/fly ash hybrid composite is observed along with hardness testing with a NHT nanoindentation tester (Anton Paar, Switzerland).

Morphological analyses have been done by scanning electron microscope (SEM, Carl Zeiss Supra 40). Sintered samples are gold-coated before performing the test through sputtering technique. Along with microscopic examination, EDS is performed.

Transmission electron microscopy (TEM, 5022/22, Tecnai G220 S-Twin, Czech Republic) is used to investigate images and SAD pattern of raw HCFC slag powder and sintered HCFC slag/fly ash hybrid composite. TEM specimen of HCFC slag is prepared by drop casting technique. TEM specimen of sintered HCFC slag is prepared by microtone technique (Leica Microsystems GmbH, A-1170). Then, specimens are transferred to carbon-coated Cu TEM grids.

FTIR analyses is conducted on fly ash (FA), HCFC slag, kaolin, feldspar, pyrophyllite, ball clay, and sintered FA-based hybrid materials. For this test, a Thermo Nicolet Nexus 870 spectrophotometer (range 400-4000 cm^{-1}) is used. Settings parameters (i.e., 50 scan at 4 cm^{-1} resolution, absorbance measurement mode) are constant and fixed in the spectrophotometers. For this test, pellet sample is prepared using KBr and test sample (small amounts) through compression molding system and pelletizer. A background spectrum is collected before the test. Then, pellet is put in a FTIR sample holder and data are collected.

Electrical resistance value of sintered pellet has been measured using two-probe method. Resistivity value pellet sample is estimated from the measured value of resistance using the following equation [31]:

$$\rho = \left(\frac{R \times A}{l}\right) \tag{9.1}$$

Here, A is the area of electrode $= \frac{\pi}{4} D^2$ (m^2), R is the resistance of prepared samples (GΩ), l is thickness of the sample, and ρ is the resistivity of the samples in Ω.cm.

9.2.4 Results and Discussion

Resistivity values of FA-HCFC hybrid composites are shown in Table 9.3. From the estimated resistivity data, the highest resistivity value (resistance value) is found to be 521.99×10^9 (36.93 GΩ) for FA (60%)/HCFC slag (10%)-based hybrid composite and lowest resistivity value (resistance value) is found to be 354.775×10^9 (25.1 GΩ) for FA (60%)/HCFC slag (5%)-based hybrid composite. This increased resistivity value is attributed to the formation of mullite phases, spinel phase, and chromium-based phase in higher amount in the composite [31, 35].

An optical image of the FA/HCFC slag-based hybrid composite (60% FA/10% HCFC slag-based hybrid composite) sintered at 1200 °C is displayed in Figure 9.3. It can be noticed from the figure that fused crystals have spread in the matrix. There are irregular shaped crystals having different textures, which are dispersed in the matrix. Since the contrast in the

Table 9.3 FA-HCFC hybrid composite:
$\left[\frac{A}{l} = 14.1345 \quad i.e, \text{radius} = 1.5 \text{ cm and thickness} = 0.5 \text{ cm}\right].$

S.no.	Sample ID	Resistance value (Gigaohm, GΩ)	Resistivity value (Ohm.cm)
1	T1	25.1	354.775×10^9
2	T2	27.67	391.10×10^9
3	T3	36.93	521.99×10^9

Note: T1 = FA (60%)/HCFC slag (5%)-based hybrid composite; T2 = FA (60%)/HCFC slag (7%)-based hybrid composite; and T3 = FA (60%)/HCFC slag (10%)-based hybrid composite.

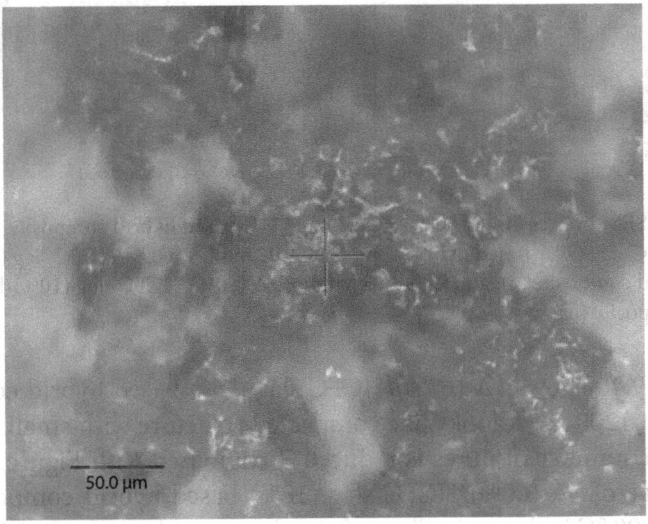

Figure 9.3 Optical image of FA/HCFC slag-based hybrid composite (60% FA/10% HCFC slag-based hybrid composite).

optical micrograph is so poor, an SEM study is carried out to understand these features in detail.

The SEM images of FA/HCFC slag-based hybrid composite sintered at 1200 °C are shown in Figure 9.4(a–c). It can be seen from the SEM image of FA (60%)/HCFC slag (5%)-based sintered hybrid composite (Figure 9.4a) that microstructure shows fused mass with rough texture. There are some void spaces in between fused masses. Phases are presumably of mixed type containing glassy phase and crystalline phase.

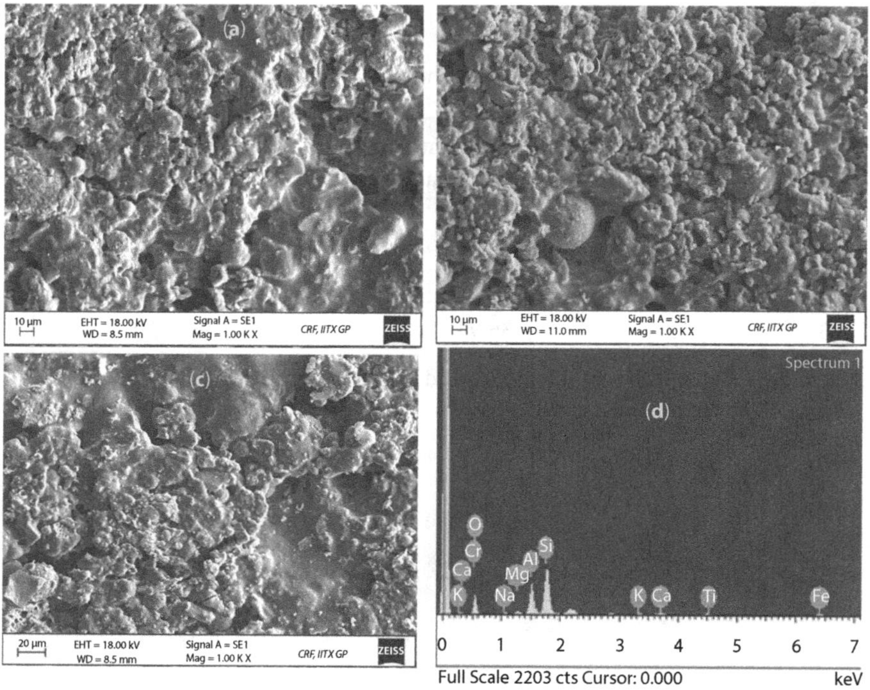

Figure 9.4 SEM images of FA (60%)/HCFC slag (5%)-based hybrid composite (a); FA (60%)/HCFC slag (7%)-based hybrid composite (b); and FA (60%)/HCFC slag (10%)-based hybrid composite (c); and EDS spectrum of FA (60%)/HCFC slag (10%)-based hybrid composite (d).

Figure 9.4b shows FA (60%)/HCFC slag (7%)-based hybrid composite sintered at 1200 °C. It looks like compacted structure with small irregular pores. The number of pores is reduced if compared with Figure 9.4a. An SEM image of FA (60%)/HCFC slag (10%)-based hybrid composite sintered at 1200 °C is shown in Figure 9.4c. Compacted structures with only a few small irregular pores are found in Figure 9.4c. This may happen due to sintering reaction occurring between raw constituents.

Figure 9.4d indicates EDS spectrum of FA (60%)/HCFC slag (10%)-based hybrid composite sintered at 1200 °C. The presence of elements, such as Si, Al, Cr, Ti, K, Ca, Cr, Fe, etc., are identified from Figure 9.4d. The EDS results are well-supported by XRD results.

A comparison is made between sintered FA (60%)/HCFC slag (10%)-based hybrid composite sintered at 1200 °C and fractured surface of FA (60%)/HCFC slag (10%)-based hybrid composite sintered at 1200 °C. Sintered sample (Figure 9.5a) shows irregular fused and crystalline masses with pores in between. Upon fracturing (Figure 9.5b), fused mass is revealed with some crystalline phase, indicating brittleness of the materials. This is due to the presence of glassy phase, which upon rupturing exposed cleavage surfaces. Along with the glassy phase, there is crystalline phase.

Figure 9.6 shows the TEM results of FA (60%)/HCFC slag (10%)-based hybrid composite sintered at 1200 °C. Figure 9.6a shows a TEM image of FA (60%)/HCFC slag (10%)-based hybrid composite sintered at 1200 °C (lower magnification), which depicts compacted spherical particles Figure 9.6b shows a TEM image of FA (60%)/HCFC slag (10%)-based hybrid composite sintered at 1200 °C at higher magnification, which looks like connected spherical particles. Figure 9.6c displays the lattice pattern of the sintered product. Lattice planes are clearly observed in the background (Figure 9.6c). Figure 9.6d shows selected area diffraction (SAD) pattern of the hybrid composite. Bright spots are observed in the SAD pattern, which is an indication of the crystalline nature of the material. This is corroborated with the XRD pattern (Figure 9.8).

The FTIR absorption bands of NALCO fly ash is shown in Figure 9.7a. The presence of different absorption stretching frequencies of fly-ash powder correspond to fly-ash components. The stretching frequency at 689 cm^{-1} matches with the Si-O-Al band. Stretching frequencies are found at 1104 and 1638 cm^{-1}. This may be accredited to Si-O-Si asymmetric band and HO-H bending band respectively [34, 36–42].

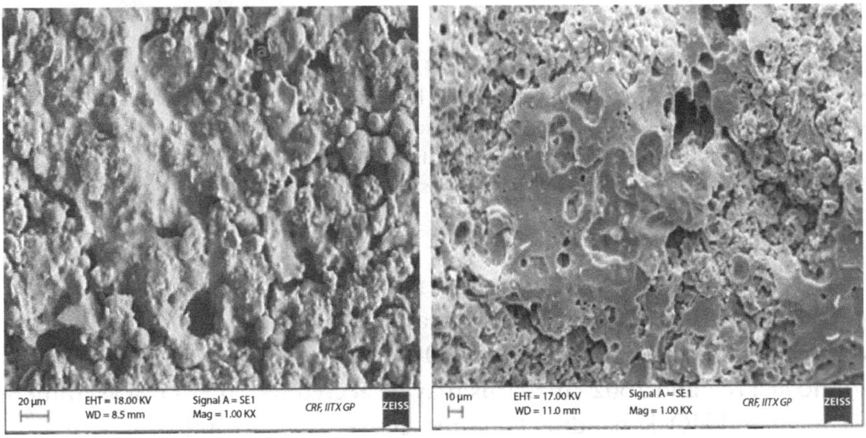

Figure 9.5 SEM image of FA (60%)/HCFC slag (10%)-based hybrid composite sintered at 1200 °C (a); and factograph of FA (60%)/HCFC slag (10%)-based hybrid composite sintered at 1200 °C (b).

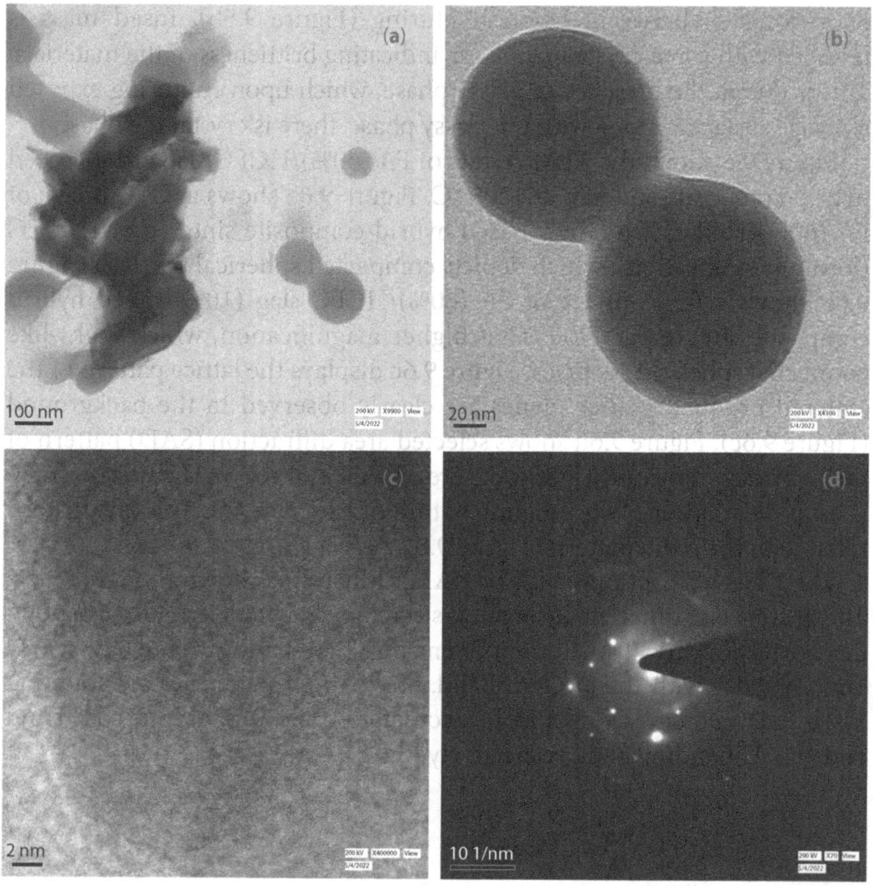

Figure 9.6 (a) TEM image of FA (60%)/HCFC slag (10%)-based hybrid composite sintered at 1200 °C (lower magnification); (b) FA (60%)/HCFC slag (10%)-based hybrid composite sintered at 1200 °C (higher magnification); (c) Lattice pattern of FA (60%)/HCFC slag (10%)-based hybrid composite sintered at 1200 °C; (d) SAED pattern of FA (60%)/HCFC slag (10%)-based hybrid composite sintered at 1200 °C.

The FTIR absorption bands of fly ash sintered at 1200 °C are shown in Figure 9.7b. The presence of different absorption stretching frequencies of fly-ash powder corresponds to fly-ash components. The stretching frequency at 559 cm^{-1} matches with Si-O-Al band. Stretching frequencies are found at 1187 and 1602 cm^{-1}. This may be accredited to Si-O-Si asymmetric band and HO-H bending band respectively [34, 36–42].

The FTIR spectrum of raw HCFC slag is shown in Figure 9.7c. From Figure 9.7c, broad stretching bands are found around 3498 cm^{-1}. The band arises due to the stretching vibrations of O-H bonds and H bending vibrations of H-O-H of interlayer adsorbed H_2O molecule. The C=O stretching

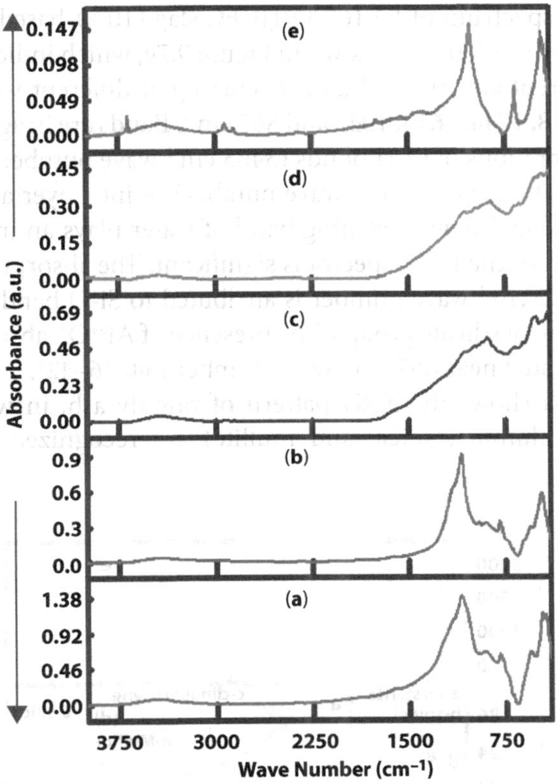

Figure 9.7 The FTIR spectra of raw FA (a), FA sintered at 1200 °C (b), raw HCFC slag (c), HCFC slag sintered at 1200 °C (d), and FA (60%)/HCFC slag (10%)-based hybrid composite sintered at 1200 °C (e).

band is indicated at 1412 cm^{-1} and confirms the presence of carbonate groups, the main reason for this being the presence of chemisorbed CO_2. Also, Si-O and O-Si-O bands are found at 1069 cm^{-1} and 567 cm^{-1}, which indicates the presence of silicate groups. The presence of Al^{3+}-O^{2-} bonds are also observed near 808 cm^{-1}; and Fe-O stretching band is also observed at 453 cm^{-1} [34, 36–42].

The FTIR spectrum of HCFC slag sintered at 1200 °C is shown in Figure 9.7d. Peak at 1100 cm^{-1} is attributed to Si–O band in SiO_4 molecules. A weak Si–O symmetrically stretching band is observed at 618 cm^{-1}. Peaks at 1100 cm^{-1} and 808 cm^{-1} are assigned as alternating Si–O and Al–O bonds. Al–O stretching vibration is present in AlO_4 tetrahedra. There is shifting of stretching frequencies. It is observed that band of asymmetric bending of O–Si–O and O–Al–O band shifts to lower value, thus confirming the previous finding [34, 36–42].

The FTIR spectrum of FA (60%)/HCFC slag (10%)-based hybrid composite sintered at 1200 °C is shown in Figure 9.7e, which indicates the presence of different absorption bands occurring at different wave numbers, i.e., 3463, 2918, 1744, 1614, 1022 and 867 cm^{-1}. Bands are in agreement with stretching vibrations of O-H bonds (3463 cm^{-1} wave number) and H-O-H bending vibrations (1614 cm^{-1} wave number) of interlayer adsorbed H_2O molecule. The hydroxyl-stretching band of water plays an important role and peak shift of the FTIR spectra is significant. The absorption band that ensues at 1022 cm^{-1} wave number is attributed to Si-O band and signifies the occurrence of silicate groups. The presence of $Al^{3+}O^{2-}$ absorption bands are also indicated near 867 cm^{-1} wave number [34, 36–42].

Figure 9.8a shows the XRD pattern of raw fly ash, in which various phases (i.e., alumina, silica, and mullite) are recognized. The highest

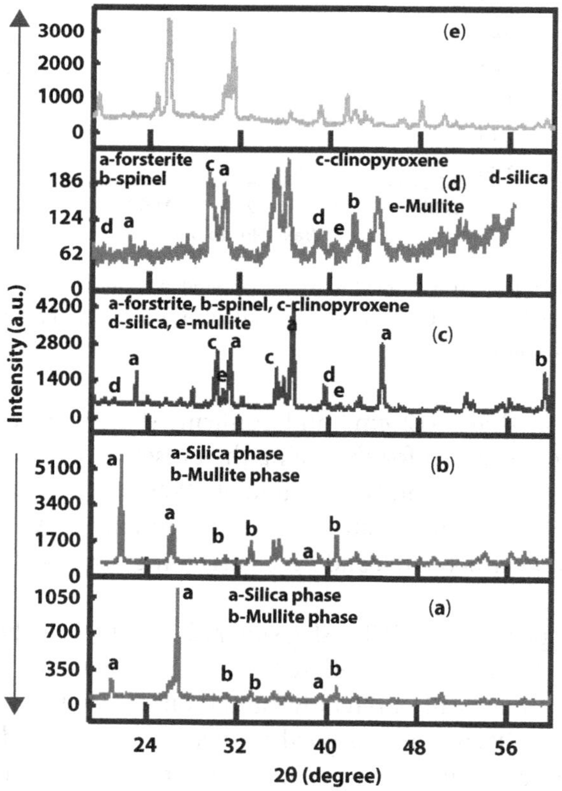

Figure 9.8 The XRD results of raw FA (a); FA sintered at 1200 °C (b); raw HCFC slag (c), HCFC slag sintered at 1200 °C (d); and FA (60%)/HCFC slag (10%)-based hybrid composite sintered at 1200 °C (e).

Table 9.4 Vickers hardness value of FA-HCFC hybrid composite sintered at 1200 °C.

S.no.	Sample ID	$F_{max.}$ (mN)	HIT (MPa)	HVIT	EIT (MPa)	E* (GPa)
1	T1	98.42	540.52	50.052	21.62	23.758

Note: T1 = FA (60%)/HCFC slag (10%)-based hybrid composite

intensity peak is identified to be silica. Mullite phase and alumina are also identified as minor quantities [43].

Figure 9.8b shows the XRD pattern of fly ash sintered at 1200 °C. Various peaks are observed in the spectrum. Mullite phase and alumina phase are present. There is marked difference in the intensity height of peaks if compared with XRD pattern of raw fly ash (Figure 9.8a), which is due to the sintering effect on fly ash at 1200 °C.

Figure 9.8c shows the XRD pattern of raw HCFC slag. Various peaks are observed in the spectrum. The most intense peak is found at 36° (2Θ) in the XRD pattern (Figure 9.8c) and the phase is clinopyroxene (ICSD analysis). Other phases, such as forsterite, spinel, mullite and silica, are also present.

Figure 9.8d displays the XRD pattern of HCFC slag sintered at 1200 °C. The most intense peak is found between 28° and 32° (2Θ) and the phase is clinopyroxene (ICSD analysis). Other phases, such as forsterite, spinel, mullite and silica, are present. There is a variation in intensities between Figure 9.8c and 9.8d, which is due to the sintering effect on HCFC slag.

Figure 9.8e displays the XRD pattern of FA (60%)/HCFC slag (10%)-based hybrid composite sintered at 1200 °C. The most intense peak is found at 27° (2Θ) in the XRD pattern (Figure 9.8e) and the phase is silica (ICSD analysis). Other phases, such as forsterite, spinel, mullite and silica, may be present. There is a large variation in intensities if compared with Figure 9.8a–d, which is due to the sintering effect of used components.

Table 9.4 shows The Vickers hardness value of FA-HCFC hybrid composite sintered at 1200 °C. HVIT value estimated through nanoindentation is 50.052. This is because of ordered bonding between atoms in the composite [45].

9.3 Conclusions

- FA/HCFC slag/CC/PY/ k-FD/BC hybrid materials were successfully prepared through sintering process (i.e., 1200 °C).

- Maximum electrical resistance value (resistivity value) of hybrid materials processed at 1200 °C was found to be 36.93 GΩ (521.99×10^9 Ω.cm).
- Mechanical property, such as Vickers hardness value of sintered hybrid sample, was found to be 50.052.
- Microstructure of fly ash showed distinct irregular spheres, whereas fused mass of crystalline phase and glassy phase were observed in fired hybrid samples.
- Optical micrograph of the highest resistivity hybrid sample showed particles with different sizes with white patches spread over the matrix.
- TEM image of the highest resistivity sintered hybrid product displayed fused mass crystals. In the SAD pattern, bright spots of sintered composite product indicated crystalline nature.
- Together with microstructural analyses, EDS analyses showed the presence of elements (Si, Al, Cr, Mg, Fe, etc.) in the fired hybrid material.
- Mullite and other phases were observed in XRD pattern.
- FTIR spectra indicated noticeable features. The corresponding groups of the peaks indicated the formation of new bonds within the phases.

Acknowledgements

The authors wish to convey their sincere thanks to the management of GIET, University Gunupur, Rayagada, Odisha, India for providing lab facilities in which to do the research work. They would also like to thank CRF IIT Kharagpur for providing testing facilities. The authors would like to thank Prof. Munesh Chandra Adhikary, PG Council Chairman, Fokir Mohan University, for his invaluable guidance, advice, and constant inspiration throughout the entire process. They would also like to thank Mr. Mukteswar Mohapatra of Fokir Mohan University for his support.

References

1. S. Babel and T. A. Kurniawan, (2003), Low Cost Adsorbents for Heavy Metals Uptake from Contaminated Water: A Review, Journal of Hazardous Materials, Vol. 97, pp. 219-243.

2. P. S. Nayak and B. K. Singh, (2007), Instrumental Characterization of Clay by XRF, XRD and FTIR," Bulletin of Materials Science, Vol. 30, pp. 235-240.
3. L. Benco, D. Tunega, J. Hafner and H. Lischka, (2001), Ab initio Density Functional Theory Applied to the Structure and Proton Dynamics of Clays, Chemistry Physics Letter, Vol. 333, pp. 479-484.
4. J. C.Hower, C. L. Senior, E. M. Suuberg, R. H.Hurt, J. L.Wilcox, and E. S. Olson, (2010), Mercury Capture by Native Fly Ash Carbons in Coal-Fired Power Plants, Progress in Energy and Combustion Science, Vol. 36, pp. 510–529.
5. S. Singh, L. C. Ram, R. E. Masto, and S. K. Verma, (2011), Acomparative Evaluation of Minerals and Trace Elements in the Ashes from Lignite, Coal Refuse, and Biomass Fired Power Plants, International Journal of Coal Geology, Vol. 87, pp. 112–120.
6. H. B. Vuthaluru and D. French, (2008), Ash Chemistry and Mineralogy of an Indonesian Coal during Combustion: Part 1 Drop-Tube Observations, Fuel Processing Technology, Vol. 89, pp. 595–607.
7. F. E. Huggins, (2002), Overview of Analytical Methods for Inorganic Constituents in Coal, International Journal of Coal Geology, Vol. 50, pp. 169–214.
8. K. Prakash and A. Sridharan, (2009) Beneficial Properties of Coal Ashes and Effective Solid Waste Management, Practice Periodical of Hazardous, Toxic and Radioactive, Waste Management, Vol. 13, pp. 239–248.
9. L. C. Ramand R. E. Masto, (2010), An Appraisal of the Potential Use of Fly Ash for Reclaiming Coal Mine Spoil, Journal of Environmental Management, Vol. 91, pp. 603–617.
10. F. S. Depoi, D. Pozebon, and W. D. Kalkreuth, "Chemical characterization of feed coals and combustion-by-products from Brazilian power plants," International Journal of Coal Geology, vol. 76, no. 3, pp. 227–236, 2008.
11. J. Levandowski and W. Kalkreuth, "Chemical and petrographical characterization of feed coal, fly ash and bottom ash from the Figueira Power Plant, Paraná, Brazil," International Journal of Coal Geology, vol. 77, no. 3-4, pp. 269–281, 2009.
12. S. Dai, L. Zhao, S. Peng *et al.*, "Abundances and distribution of minerals and elements in high-alumina coal fly ash from the Jungar Power Plant, InnerMongolia, China," International Journal of Coal Geology, vol. 81, no. 4, pp. 320–332, 2010.
13. J. C. Egbai, (2013), Kaolin Quantification in Ukwu-nzu and Ubulu-uku using Electrical Resistivity Method, International Journal of Recent Research and Applied Studies, Vol. 14, pp. xxx-xxx.
14. O. A. Babalola and T. Akomolafe, (2008), Effects of Kaolin Particle Size and Annealing Temperature on the Resistivity of Zinc-Kaolin Composite Resistors, Journal Applied Science Environmental Management, Vol. 12, p. 113-xxx.
15. L. Li and Y. Li, (2017), Development and Trend of Ceramic Cutting Tools from the Perspective of Mechanical Processing, IOP Conf. Series: Earth and Environmental Science, Vol. 94, pp.1-6.

16. M. Zhu, R. Ji, Z. Li, H. Wang, L. L. Liu, and Z. Zhang, (2016), Preparation of Glass Ceramic Foams for Thermal Insulation Applications from Coal Fly Ash and Waste Glass, Construction and Building Materials, Vol. 112, pp. 398-405.
17. H. E. Exner and E. Arzt, (1996), Sintering Processes, In Physical Metallurgy, Ed. R. W. Cahn and P. Haasen, 4th Ed. Elsevier Science, Amsterdam, Vol.3, pp. 2628-2662.
18. W. V. Siemens, Inventor and Entrepreneur: Recollections of Werner Von Siemens, London, England, 1966.
19. T. Tanaka and T. Imai, (2013), Advances in Nanodielectric Materials Over the Past 50 years, IEEE Electrical Insulation Magazine, Vol. 1, pp. 10-23.
20. T. Tanaka, (2005), Dielectric Nanocomposites with Insulating Properties, IEEE Transactions on Dielectrics and Electrical Insulation, Vol. 12, pp. 914-928.
21. https://www.vedantu.com/evs/uses-of-insulators
22. https://physics.stackexchange.com/questions/172076/how-does-insulating-footwear-prevent-an-electric-shock
23. https://ohsonline.com/articles/2021/09/15/six-steps.aspx
24. J. Martín-Márquez, J. Ma. Rincón and M. Romero, (2008), Effect of firing temperature on sintering of porcelain stoneware tiles, Ceramics International, Vol. 34, pp. 1867-1873; (doi:10.1016/j.ceramint.2007.06.006).
25. https://www3.epa.gov/ttnchie1/ap42/ch11/final/c11s07.pdf
26. K. Galos, (2011), Composition and ceramic properties of ball clays for porcelain stoneware tiles manufacture in Poland, Applied Clay Science, Vol. 51, pp. 74-85.
27. https://www.nrdc.org/onearth/coal-ash-fly-ash-bottom-ash-and-boiler-slag
28. Y. Luo, S. Zheng, S. Ma, C. Liu, and W. Xiaohui, (2017), Ceramic tiles derived from coal fly ash: Preparation and mechanical characterization, Ceramics International, Vol. 43, pp. xxx-xxx (DOI: 10.1016/j.ceramint.2017.06.045).
29. M. K.Panigrah, P. Kumar, B. Barik, S. K. Mohapatra, H. Jha, R. R. Dash, Frequency Dependency of Developed Dielectric Material from Fly Ash: An Industrial waste" in 20th National Conference on Non-Ferrous Minerals and Metals-2016
30. M. Panigrahi, R. I. Ganguly, R. R. Dash, (2018), An Advanced Cured High Carbon Ferrochrome Slag (HCFCS) Geopolymer (GP): A Constructional Materials IOP Conference Series: Materials Science and Engineering, Vol. 410, pp.1-15
31. M.K. Panigrahi, (2021), Investigation of Structural, Morphological, Resistivity of Novel Electrical Insulator: Industrial Wastes, Bulletin of Scientific Research, Vol. 3, pp. 51-58
32. X. Liu and M. Bai, (2017), Effect of Chemical Composition on the Surface Charge Property and Flotation Behavior of Pyrophyllite Particles, Advanced Powder Technology, Vol. 28, pp. 836–841.

33. X. Liu, and M. Bai, (2011), Composition and Ceramic Properties of Ball Clays for Porcelain Stoneware Tiles Manufacture in Poland, Applied Clay Science, Vol. 51, pp. 74–85.
34. N F Ayoob, J M. Juoi, Z.M. Rosil, and N R. Rosil, Characterization and properties of sintered glass-ceramic produced from Recycling of glass by using pressure-less method, Key Engineering Materials Vols. 471-472 (2011) pp 933-938 (doi:10.4028/www.scientific.net/KEM.471-472.933)
35. M.K. Panigrahi, (2022), Investigation of Structures of Sintered Fly Ash Materials: Resources of Industrial Wastes, Bulletin of Scientific Research Vol. 4 pp.1-10.
36. B.J. Saikia, G.K Rao, and P. Sarathy, (2010), Fourier Transform Infrared Spectroscopic Characterization of Kaolinite from Assam and Meghalaya, North Eastern India, International Journal of Modern Physcs, Vol. 1, pp. 206-xxx.
37. R.P. dos Santos, J. Martins, C. Gadelha, B. Cavada, A.V. Albertini, F. Arruda, M. Vasconcelos, E. Teixeira, F. Alves, J.L. Filho, and V. Freire, (2014), Coal Fly Ash Ceramics: Preparation, Characterization, and Use in the Hydrolysis of Sucrose. Scientific World Journal, Vol. 2014, pp. 01.
38. A. Jose, M. R. Nivitha, M. Krishnan, and R. G. Robinson, (2020), Characterization of Cement Stabilized Pond Ash using FTIR Spectroscopy, Construction and Building Materials, Vol. 263, pp.120136-xxx (DOI: 10.1016/j.conbuildmat.2020.120136).
39. Y. Liu, F. Zeng, B. Sun, P. Jia, and I. T. Graham, (2019), Structural Characterizations of Aluminosilicates in Two Types of Fly Ash Samples from Shanxi Province, North China, Minerals, Vol. 9, pp. 358-16 (doi:10.3390/min9060358).
40. P. Rożek, M. Król, and W. Mozgawa, (2018), Spectroscopic Studies of Fly Ash-Based Geopolymers, Spectrochim Acta - Part A Molecule, Biomolecule Spectroscopy Vol. 198, pp. 283-289 (https://doi.org/10.1016/j.saa.2018.03.034).
41. S. Kumar, F. Kristály, and G. Mucsi, (2015), Geopolymerization Behaviour of Size Fractioned Fly Ash, Advanced Powder Technology, Vol. 26, pp. 24-30 (https://doi.org/10.1016/j.apt.2014.09.001).
42. Z. Kledyński, A. Machowska, B. Pacewska and I. Wilińska, (2017), Investigation of Hydration Products of Fly Ash-Slag Pastes, Journal of Thermal Analysis and Calorimetry, Vol. 130, pp. 351-363 (https://doi.org/10.1007/s10973-017-6233-4).
43. M. Panigrahi, R. R. Dash, R. I. Ganguly, (2018), Development of Novel Constructional Material from Industrial Solid Waste as Geopolymer for Future Engineers IOP Conference Series: Materials Science and Engineering, Vol. 410, pp.1-12

10

High Resistant HCFC Slag/ Pond Ash Hybrid Ceramics

M. K. Panigrahi[1]*, R.I. Ganguly[2] and R.R. Dash[3]

[1]*Department of Materials Science, Maharaja Sriram Chandra Bhanja Deo University, Keonjhar Campus, Odisha, India*
[2]*Department of Metallurgical Engineering, National Institute of Technology, Rourkela, Odisha, India*
[3]*National Metallurgical Laboratory, Jamshedpur, Jharkhand, India*

Abstract
Coal ash is a by-product of coal combustion in thermal power plants. Since ash is highly accountable for environmental pollution, it is essential to find novel methods of reusing it.

This chapter discusses the processing of hybrid materials with solid-state sintering. A hybrid ceramic material is developed by mixing and sintering materials such as pond ash (PA), high-carbon ferrochrome (HCFC) slag, kaolin (CC), k-feldspar (k-FD), pyrophyllite (PY), and ball clay (BC). Such hybrid material is prepared using sintering temperatures of 1200 °C, and after calculating the electrical resistivity of the fired material, the highest resistivity achieved was determined to be 23.71 GΩ (335×10^9 Ω.cm) for batch consisting of PA (60%)/HCFC slag (10%)/CC (10%)/PY (5%)/k-FD (10%)/BC (5%) hybrid composite. Characterizations of developed materials are done with X-ray diffraction (XRD), Fourier-transform infrared spectroscopy (FTIR), optical microscopy, field emission scanning electron microscopy (FESEM) with EDS and transmission electron microscopy (TEM). These XRD and FTIR studies have enabled the identification of the structure and chemical group of the developed material. Surface topography of hybrid samples has also been analyzed. Among the phases identified, mullite is found to be the dominating phase in the sintered material, since it enhances the insulation property of the developed material.

**Corresponding author*: muktikanta2@gmail.com

Muktikanta Panigrahi, Ratan Indu Ganguly and Radha Raman Dash (eds.) High Electrical Resistant Materials: Ferrochrome Slag Resource Ceramics, (217–240) © 2024 Scrivener Publishing LLC

***Keywords*:** Pond ash, kaolin, pyrophyllite, k-feldspar, ball clays, mullite phases, insulator

10.1 Introduction

Thermal power plants generate large quantities of coal ash [1]. Handling large quantities of coal ash has become a problem. This is a waste product which is usually dumped in a landfill, pit or pond [2, 3]. Therefore, cultivable land is wasted. The quantity of ash generated by a thermal power plant depends on the source, i.e., composition of coal. The physicochemical properties of coal ash differ from each other [4]. This is because the source of coal procurement is not the same. Coal procured from the same mines will differ in composition depending on the depth at which it is collected. Usually, coal ash contains carbon, sulfur, silica, alumina, iron oxide, calcia, magnesia, etc. [5]. Usually, coal ash is categorized into three types: fly ash, bottom ash, and pond ash [6]. Fly ash is one of the by-products that is collected by electrostatic precipitators, mechanical filters, or is simply ejected into the atmosphere [7]. It is composed of hollow spherical particles known as cenospheres. They are hollow spherical particles with low density. Cenosphere material consists of glassy aluminum-silicate matrix with other phases, i.e., quartz, mullite, calcite, and magnetite [8]. Usually, 70% fly ash is generated by thermal power plants during the combustion of coal [9]. The presence of fly ash pollutes the environment and is highly harmful to living beings. Also, fly ash pollutes soil, water, and air environments [10].

The ashes left behind after burning coal is dumped into ponds and settles at the bottom of the pond. They pollute water bodies and cause harm to aquatic animals. With increasing power requirement, more and more coal is being burnt and pond ash is being generated at an alarming rate. Thermal power plant management is now facing the difficulty of its disposal. Hence, utilization of wastes has become a challenge for engineering researchers [11].

Because of its composition, coal ash has potential use as additive in many technological applications, i.e., manufacture of cement and concrete, conventional ceramics, and vitroceramics, among others [12].

During the last decades, ceramics have been used in different sectors such as refractories [13], cutting tools [14], thermal insulations [15], etc. Ceramics are prepared by pressure and firing temperature. Qualities of ceramic products are evaluated from their microstructural constituents i.e., crystallized phase(s). The firing process plays a vital role in the development of crystallized phases in the materials. Formation of mullite phase enhances properties. This phase is usually found by sintering in the temperature range of 1150–1300 °C [16].

In 1850, ceramic materials were employed as electrical insulators during the construction of electrical air lines [17]. Ceramics have a number of distinctive properties (i.e., mechanical strength, high-power dielectric strength, and corrosion resistance) [18, 19]. Currently, low-cost insulators with high efficiency are needed in areas of the electrical engineering sector. Current demands have motivated researchers to develop insulators with the best properties to meet our requirements [20]. Insulators have great importance in day-to-day life and make our lives easier, safer and shock-free [21–23].

Sintered ceramics have water absorption and open porosity property close to zero [24, 25]. This is why they possess the perfect physico-mechanical properties, i.e., high hardness, abrasion resistance and bending strength. These features of sintered ceramics are due to the mineralogical composition of the raw materials, i.e., kaolin, feldspar quartz, pure quartz sand, and ball clays. Also, some wastes, such as fly ash, pond ash, bottom ash, slags, etc., are used to prepare ceramics. Recently, researchers have put their effort into exploring cost-efficient ceramics made from different mineralogical compositions.

Kaolinite is an essential low-cost and highly available aluminosilicate clay mineral. It has weathering, diagentic, hydrothermal and very low-grade metamorphic features. It mostly occurs as clay-sized particles and has high aspect ratios. Therefore, kaolinite may play a dynamic role in controlling the properties of materials (i.e., degree of crystallinity, concentration of impurities, particles' size distribution). Clays are widely employed in different sectors [26].

Pyrophyllite is a hydrous layered aluminum silicate [$Al_2Si_4O_{10}(OH)_2$]. It has 2:1 structure. Individual sheets of pyrophyllite consist of $Al(O, OH)_6$ octahedra sandwiched between sheets of SiO_4 tetrahedra. The components of pyrophyllite are SiO_2 (67%), Al_2O_3 (28%) and H_2O (5%) [27–29]. Thus, impurities in pyrophyllite ore are generally quartz, feldspar, diaspore, dickite, chlorite and mica [30–32]. Pyrophyllite has potential use in various areas such as ceramic, plastic, rubber, pressure transmission medium, medical carrier and adsorption or photocatalytic materials. This applicability arises due to its good physico-chemical characteristics (i.e., low thermal and electrical conductivity, low expansion coefficient, low reversible thermal expansion and excellent reheating stability) [28, 31, 33–36]. Many studies are available in the literature on the synthesis of clay mineral-based nanocomposites [36–48]. Pyrophyllite is also one of the main gangue silicates in diasporic bauxite [49–51] and is recovered by flotation process. In the process, characteristics of minerals (i.e., electrical properties, chemical composition and particle size distribution) are vital factors and are due to flotation behavior [48, 49, 52–57].

Feldspars are an important class of materials because of their well-known abundance in the Earth's upper crust, which covers 50 wt% of the Earth's crust [58]. Feldspars are the chief rock-forming minerals in magmatic, metamorphic, and sedimentary rocks. Reports on conductivity of alkali feldspar supports the investigation of the electrical properties of the feldspar family [59][a-e]. Studies on the electrical conductivity of feldspar have been carried out at room and high temperature. Literature is available on estimation of the pressure dependence of the glass transition temperature (T_g) and activation energy required for electrical conductivity in albite, haplogranite, anorthite, and SiO_2 glasses [60][a,b]. Estimation of electrical conductivity in dry and hydrous state of albite glasses and liquids at both high-pressure and high-temperature conditions is also reported [61]. Electrical conductivity has been measured for dry and hydrous plagioclase at high temperature and pressure. Samples (natural and pre-annealed) are made from fresh mafic granulite xenoliths. The effect of chemical composition on electrical conductivity of feldspar has not been quantitatively assessed. Although conduction of feldspar is described as the migration of alkali cations (i.e., Na^+, K^+ and Ca^{+2}), it is still not clear whether alkali cations are transported by a vacancy mechanism. This is also a controversial issue for alkali ions dynamics in the diffusion measurements of feldspar [62].

Ball clay is used in bulk quantities as raw material for production of porcelain stoneware and tile [63]. This material has good molding properties and high mechanical strength. Ball clay is composed of kaolinite (25–80%), illite, and mica (10–30%), fine-grained quartz, illite/smectite (I/S) mixed-layered minerals and organic matter (up to 0.5 mass %) [64][a-c]. Occurrence of kaolinite stimulates mullite formation, whereas existence of illite and smectite influences glassy phase formation, good densification and high whiteness of the ceramic body during firing [64[a-c], 65[a, b]].

Ceramic properties and level of crystallinity of ball clays depend on the content of the clay minerals, particularly illite/smectite minerals. Ball clays usually have different sizes and amounts of grains [b1 μm (50–90%), and even b0.2 μm (25–40%)]. The grain size of clay is related inversely to the specific BET surface area and plasticity. The level of crystallinity influences the plasticity of clay [66, 67]. Thus, optimum ball clays should comprise higher extents of low-ordered kaolinite, illite, and some amounts of I/S minerals. Due to these features, ball clay shows very good sintering property. Thus, ball clay is highly reactive towards other components of the ceramic batch [65[b]].

To enhance the insulation properties of pond ash, it is essential to add some other minerals during the preparation of composite. Reports are available on industrial waste-based composite [68, 69]. Usually, industrial wastes contain different types of elements (i.e., Cu, Pb, Cd, Ag, Mo, Fi, Ti, Na, Mn, S, P, Zn, and Cl) in different proportions.

Thermal power plant ash mainly consists of crystalline and glassy phase [70]. Work is going on to enhance the dielectric and electrical insulation properties of industrial wastes [71–74].

Direct current (DC) resistivity, dielectric constant, and dielectric loss are the main considerations for fabricating insulator [75].

The objective of the present work is to prepare sintered PA/HCFC slag/CC/PY/k-FD/BC hybrid composite through sintering process via solid state route. Materials are fired at 1200 °C. Microstructure and elemental analyses of sintered materials are analyzed by FESEM coupled with EDS. Different phases present in sintered material have been investigated by X-ray diffraction technique, and FTIR analyses are done to find the presence of chemical groups in the samples. Electrical resistivity value is also estimated by two-probe method.

10.2 Experiment Details

10.2.1 Materials and Chemicals

Pond ash was procured from NALCO (a Navratna CPSE under Ministry of Mines), Damanjodi, Odisha, India. Kaolin, dextrin $(C_6H_{10}O_5)_n{\cdot}xH_2O$, k-feldspar, pyrophyllite, and ball clays are procured from Merck, India. A few drops of water are required during preparation of green paste. Compositions of raw samples, such as processed pond ash, kaolin, feldspar, pyrophalite, and ball clays, are showed in Table 10.1. Optical images of raw samples (i.e., pond ash, kaolin, dextrin, k-feldspar, pyrophyllite, and ball clays) are shown in Figure 10.1.

10.2.2 PA/HCFC Slag Hybrid Materials Peparation

Preparation of PA/HCFC Slag hybrid ceramic is achieved through solid-state route followed by sintering process [78]. The following steps are used to prepare PA/HCFC slag hybrid materials [78]:

Step 1: Raw materials preparation process
Pond ash is ground in a ball mill for 5 h to reduce sizes of pond ash. It is screened through 240 meshes. Screened pond ash is dried in a heating

Table 10.1 Compositions with percentage (%) of as-processed pond ash, kaolin, pyrophyllite, k-feldspar, and ball clays.

S. ID	Compositions (%)												
	Fe_2O_3	Al_2O_3	SiO_2	P_2O_5	Cr_2O_3	ZnO	MgO	C	CaO	TiO_2	MnO	LOI	Ref
PA	3.85	28.3	62.8	0.32	0.04	0.027	0.049	1.15	0.7	1.84	0.03	0.5	[76]
HCFC SLAG	2.5-3.0	23.5-24.5	28-31	---	4.5-5.4	---	26.5-26.0	---	7.5-8.5	---	---	---	[77]
CC	---	37.66	44.80	---	---	---	T	---	0.50	0.60	---	14.33	[78]
PY	---	27.81	65.95	---	---	---	0.64	---	0.022	0.12	---	5.92	[79]
FD	---	18.18	64.57	---	0.22	---	---	---	---	0.01	----	---	[80]
BC	0.90	27.0	58.0	---	---	0.2	<0.1	---	<0.1	1.5	---	---	[81]

Note: S. ID = Sample ID; PA = Pond Ash; CC = Kaolin; PY = Pyrophyllite; k-FD = k-Feldspar (Potassium); BC = Ball clays; T = Trace.

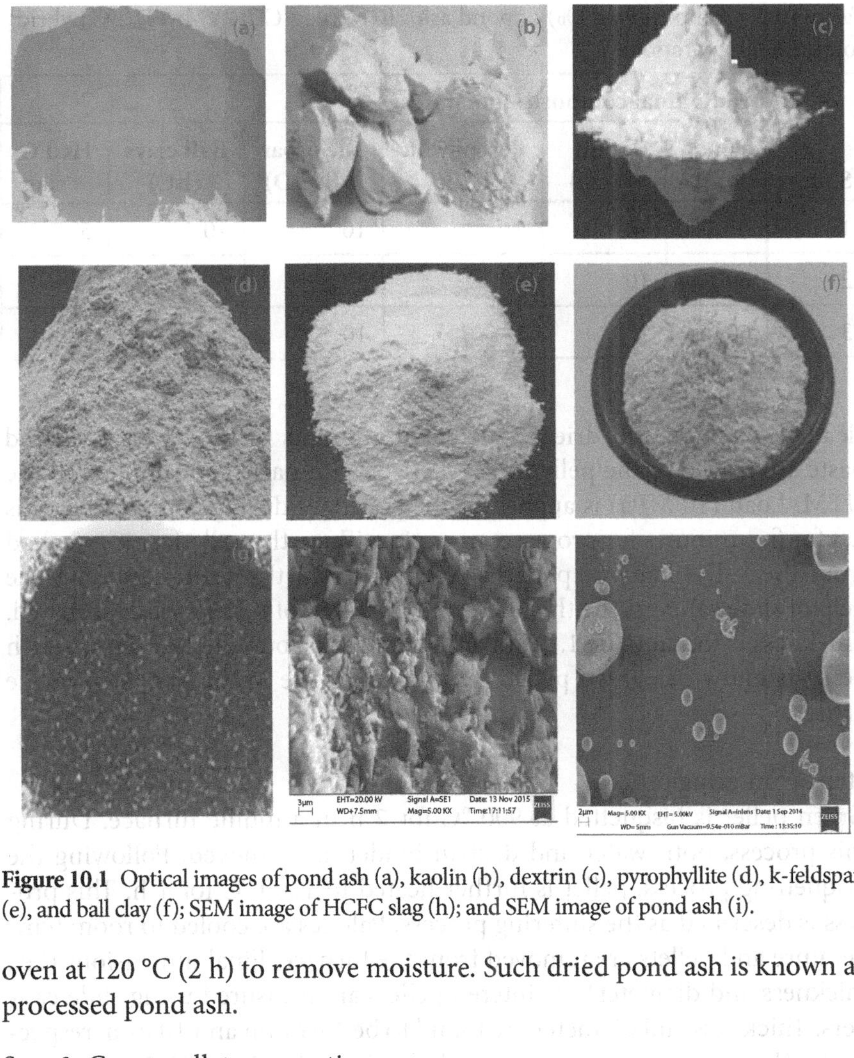

Figure 10.1 Optical images of pond ash (a), kaolin (b), dextrin (c), pyrophyllite (d), k-feldspar (e), and ball clay (f); SEM image of HCFC slag (h); and SEM image of pond ash (i).

oven at 120 °C (2 h) to remove moisture. Such dried pond ash is known as processed pond ash.

Step 2: Green pellet preparation process
Processed pond ash (60%) is put into a mortar with pestle as main component. Sieved HCFC slag, kaolin, pyrophyllite, k-feldspar (potassium), and ball clays amounting to 40% are mixed with 60% PA. Individual percentages of component are shown in Table 10.2. Total mixture is ground in a mortar with pestle for half an hour. This ensures the uniformity of the ground mixture. Dextrin (i.e., 0.05%) is added to the mixture. It is further ground for 1 h. A few drops of water are added to the mixture to make it pasty. Such paste samples are known as green samples. Pellets are made from paste.

Pellet is made using a universal testing machine (UTM) with pelletizer. Before beginning the process, different parts of the pelletizer are properly

Table 10.2 Compositions (%) of pond ash/HCFC slag/CC/PY/ k-FD/BC hybrid composite preparation.

S.no.	Individual composition name					
	Pond ash	**Kaolin (CC)**	**Pyrophyllite (PY)**	**k-feldspar (k-FD)**	**Ball clays (BC)**	**HCFC slag**
1	60	10	5	10	10	5
2	60	10	8	10	5	7
3	60	10	5	10	5	10

cleaned, washed, and dried. Calculated amounts of the above prepared paste are placed in the pelletizer. Pelletizer with paste is positioned in the UTM. Load (10 MPa) is applied to the pelletizer during compaction. It is left for five minutes for proper compaction. Then, the pelletizer is removed from the UTM and the pellet is removed from the pelletizer. With the help of slide calipers, the thickness and diameter of pellet(s) are estimated. Thickness is found to be 1.1 mm and diameter is found to be 10 mm. Such pellet is known as green pellet. It is ready for use in the next step of the process.

Step 3: Sintering process
Green pellet is bisquetted at 900 °C for 2 h in a muffle furnace. During this process, both water and dextrin binder are removed. Following the bisquetting process, pellet is further heated to 1200 °C for 1 h. This process is described as the sintering process. Palettes are cooled to room temperature and pellets are removed from the furnace. Final dimensions (i.e., thickness and diameter) of sintered pellets are measured using slide calipers. Thickness and diameter are found to be 1.09 mm and 10 mm, respectively. The entire process is shown schematically in Figure 10.2.

Sintered pellet(s) is ready for use in different tests. Other sets of reaction are carried out in a similar fashion.

10.2.3 Test Methods

X-ray diffraction (XRD) analyses are carried out using a Philips PW-1710 Advance wide-angle X-ray diffractometer coupled with a Philips PW-1729 X-ray generator to identify different phases. CuKα radiation (λ ~0.154 nm) is used for diffraction studies. The XRD machine functions at 40 kV and 20 mA. Prepared pellet sample is put on a quartz sample holder. Sample is

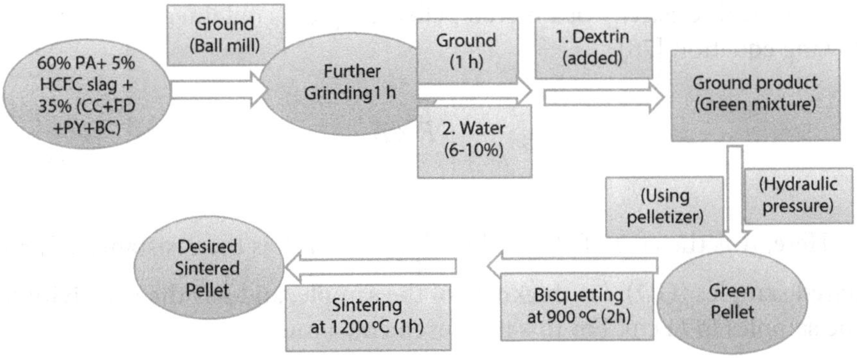

Figure 10.2 Flow chart showing the preparation of batch-1 [Pond Ash (60%) + HCFC slag (5%) + 35% others (CC+FD+PY+BC) composite sintered at 1200 °C].

scanned through diffraction angle from 10° to 70° with scanning rate of 5°/min.

Optical imaging and Vickers hardness value of sintered HCFC slag/pond ash hybrid composite are taken. A nanoindentation tester (NHT; Anton Paar, Switzerland) is used for tests.

Morphological studies are done by field emission scanning electron microscopy (FESEM, Carl Zeiss Supra 40). Test samples are gold-coated using sputtering technique. Energy-dispersive X-ray spectroscopy (EDS) is performed in the same experiment.

Transmission electron microscopy (TEM) is used to investigate images and SAD pattern of raw HCFC slag powder and sintered HCFC slag/pond ash hybrid composite. A 5022/22, Tecnai G220 S-Twin transmission electron microscope (Czech Republic) is used to carry out the measurement. The TEM specimen (HCFC slag) is prepared by drop casting technique, and TEM specimen of sintered HCFC slag is prepared by microtone technique (Leica Microsystems GmbH, A-1170). Then, test specimens are transferred to carbon-coated Cu TEM grids.

The FTIR analyses of pond ash (PA), kaolin, k-feldspar, pyrophyllite, ball clay, and sintered PA-based hybrid materials are made using a Thermo Nicolet Nexus 870 spectrophotometer (range 400-4000 cm^{-1}). Setting parameters (i.e., 50 scan at 4 cm^{-1} resolution, absorbance measurement mode) are constant and fixed with spectrophotometer. Powder samples of hybrid material are mixed with KBr. Pellets from powder sample are then prepared using a compression molding system and pelletizer. Before the test, a background spectrum is collected. Then, pellet is transferred to FTIR sample holder and data are collected.

Electrical resistivity of sintered pellet samples is obtained using the following equation [78]:

$$\rho = \left(\frac{R \times A}{l} \right) \tag{10.1}$$

Here, A is the area of electrode $= \frac{\pi}{4} D^2$ (m^2), R is the resistance of prepared samples (GΩ), l is thickness of the sample, and ρ is the resistivity of the samples in Ω.cm. For this measurement, a multimeter is used.

10.2.4 Results and Discussion

Table 10.3 shows the mechanical property (i.e., Vickers hardness value) of FA-HCFC hybrid composite sintered at 1200 °C. Estimated HVIT value and Young's modulus values arrived at through nanoindentation are 393.87 and 50.421 MPa, respectively.

Table 10.4 shows the resistance and resistivity value of three PA/HCFC slag hybrid composites designated by T1, T2, and T3 (see Table 10.2), respectively. Among the three hybrid composites, the maximum resistance and resistivity value are found to be 23.71 GΩ and 335.13×10^9 Ohm.cm, respectively. Among the three hybrid composites, T1 composition has shown the lowest resistance and resistivity value, which are 11.3 GΩ and 159.72×10^9 Ohm.cm, respectively. This is due to formation of more amount of mullite phases (T3-composition). Mullite has been found to enhance resistance and resistivity value of the material.

An optical image of PA/HCFC slag hybrid composite sintered at 1200 °C is displayed in Figure 10.3. It is noticed from Figure 10.3 that fused crystals are spread over the matrix. There are irregular shaped crystals with different aspect ratios. Optical microscopy shows poor contrast. SEM studies are carried out to understand the features in detail.

Microstructure shown in Figure 10.4a is compact mass with irregular texture. There are some voids between the compact mass. Similar features

Table 10.3 Vickers hardness value of PA-HCFC hybrid composite sintered at 1200 °C.

S.no.	Sample ID	F_{max}.(mN)	HIT (MPa)	HVIT	EIT (MPa)	E* (GPa)
2	T2	49.87	4252.9	**393.87**	50.421	55.408

Note: T3 = PA (60%)/HCFC slag (10%)-based hybrid composite; HV(IT) = Vickers hardness.

Table 10.4 Electrical resistivity of PA-HCFC hybrid composite (sintered at 1200 °C for 1 h).

S.no.	Sample ID	Resistance value (Gigaohm, GΩ)	Resistivity value (Ohm.cm)
1	T1	11.3	159.72×10^9
2	T2	13.22	186.86×10^9
3	T3	23.71	335.13×10^9

Note: T1 = PA (60%)/HCFC slag (5%)-based hybrid composite; T2 = PA (60%)/HCFC slag (7%)-based hybrid composite; and T3 = PA (60%)/HCFC slag (10%)-based hybrid composite.

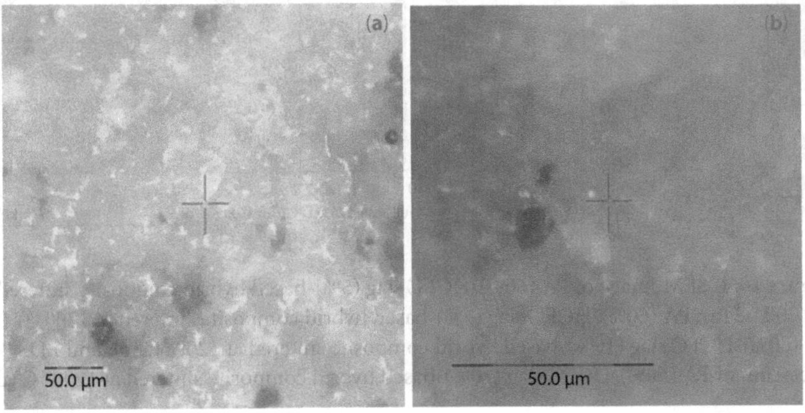

Figure 10.3 Optical images of PA sintered at 1200 °C (a) and PA (60%)/HCFC slag (10%)-based hybrid composite (b).

are observed for the microstructures shown in Figure 10.4b and 10.4c. Images are shown to contain more glassy phase. The EDS spectrum of PA (60%)/HCFC slag (10%)-based hybrid composite sintered at 1200 °C is shown in Figure 10.4d. It is observed that mass contains elements O, Si, Al Cr, Ca, Mg, Fe, Ti, K, and Na. The EDS results are well-supported by XRD results.

An SEM image of PA (60%)/HCFC slag (10%)-based hybrid composite sintered at 1200 °C is shown in Figure 10.5a. It is seen from the figure that PA particles have been fused with HCFC slag and other minor ingredients (kaolin, K-feldspar, pyrophyllite, ball clays). The compacted structure is irregularly shaped with rough texture. There are voids present in between compacted mass. Figure 10.5b shows a factograph of PA (60%)/HCFC slag

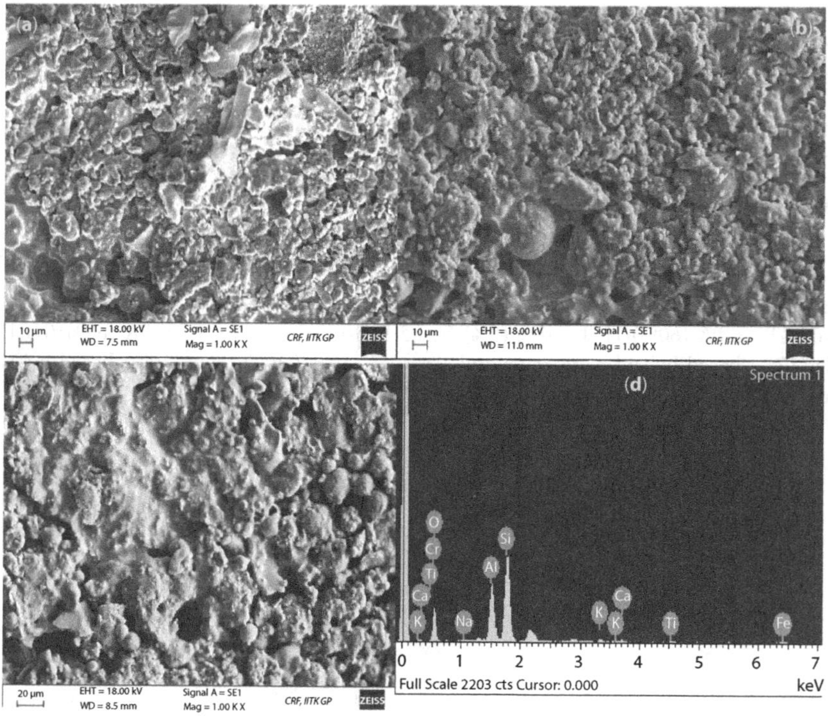

Figure 10.4 SEM image of PA (60%)/HCFC slag (5%)-based hybrid composite sintered at 1200 °C (a); PA (60%)/HCFC slag (7%)-based hybrid composite sintered at 1200 °C (b); PA (60%)/HCFC slag (10%)-based hybrid composite sintered at 1200 °C (c); and EDS Spectrum of PA (60%)/HCFC slag (10%)-based hybrid composite sintered at 1200 °C (d).

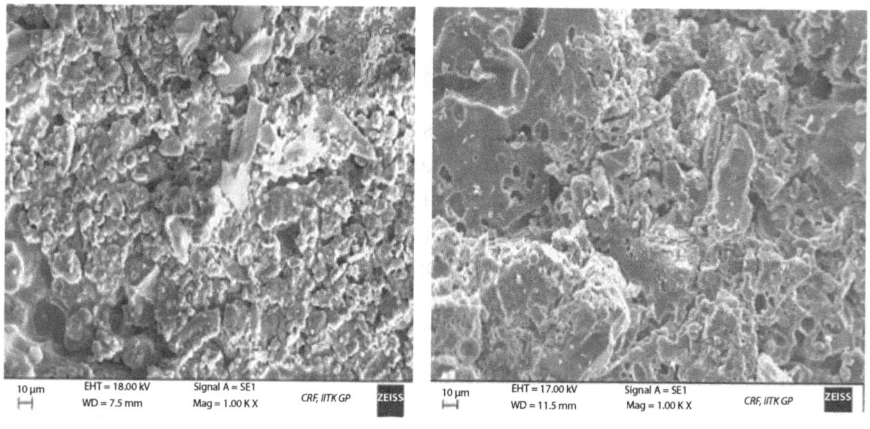

Figure 10.5 SEM image of PA (60%)/HCFC slag (10%)-based hybrid composite sintered at 1200 °C (a) and factograph of PA (60%)/HCFC slag (10%)-based hybrid composite sintered at 1200 °C (b).

(10%)-based hybrid composite sintered at 1200 °C. Figure 10.5b indicates some cleavage faces containing some pores. In some places, faces having smaller surface area are seen. There are holes present in the structure.

Figure 10.6a shows the TEM image of PA (60%)/HCFC slag (10%)-based hybrid composite sintered at 1200 °C (lower magnification). In the TEM image (Figure 10.6a), crystalline structures are observed. In the higher magnification (Figure 10.6b), crystalline structure with different shapes is clearly indicated. Figure 10.6c indicates different sized bright spots spread in the background. This indicates the crystalline nature of the prepared

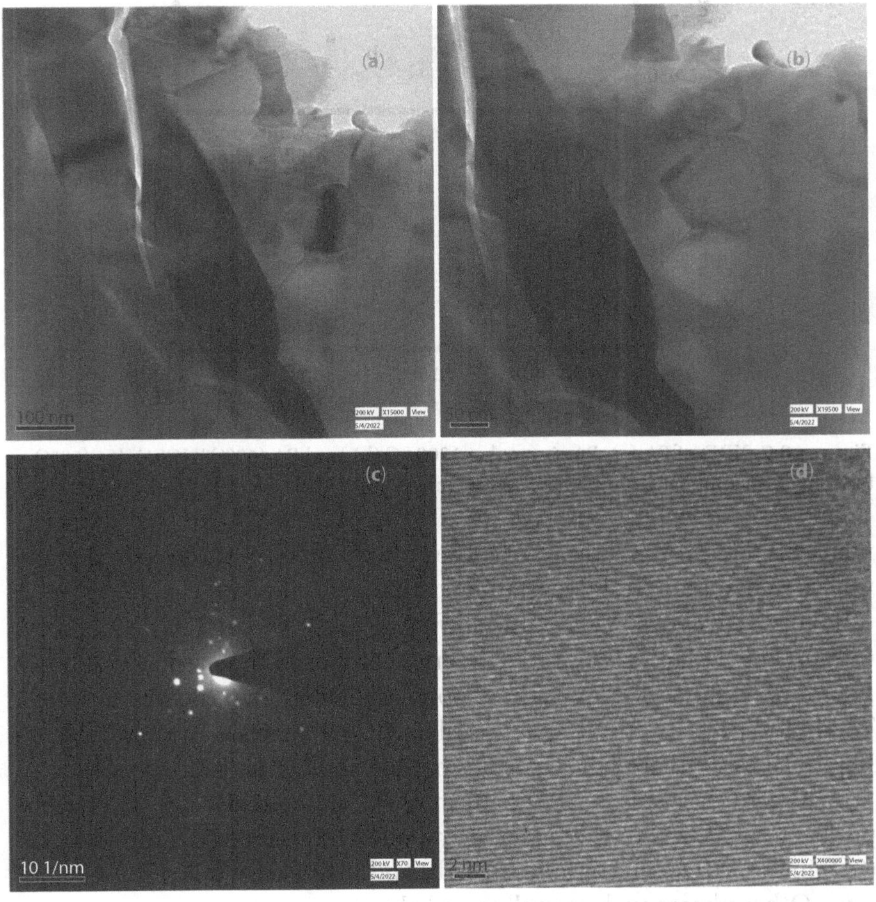

Figure 10.6 (a) TEM analyses of PA (60%)/HCFC slag (10%)-based hybrid composite sintered at 1200 °C (lower magnification); (b) TEM analyses of PA (60%)/HCFC slag (10%)-based hybrid composite sintered at 1200 °C (higher magnification); (c) SAED pattern of PA (60%)/HCFC slag (10%)-based hybrid composite sintered at 1200 °C; (d) Lattice pattern of PA (60%)/HPCF slag (10%)-based hybrid composite sintered at 1200 °C.

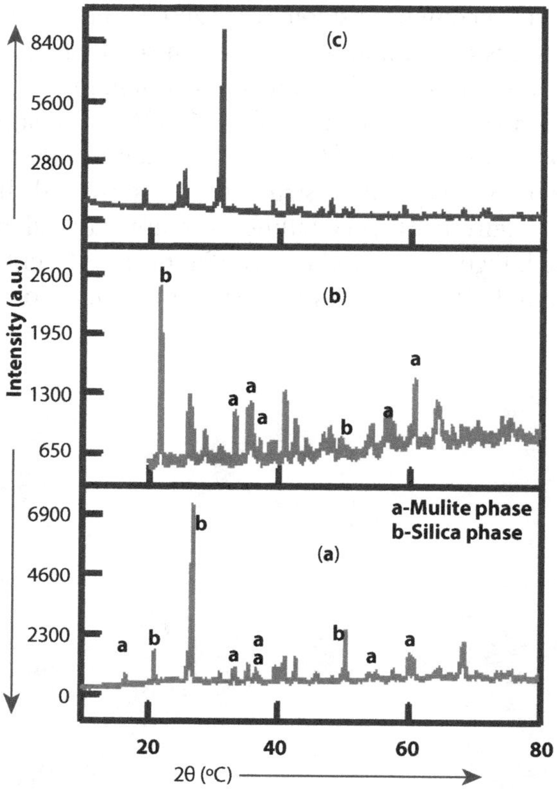

Figure 10.7 XRD of PA (a), PA sintered at 1200 °C (b), and PA (60%)/HCFC slag (10%)-based hybrid composite sintered at 1200 °C (c).

materials. Figure 10.6d shows that the lattice pattern of the prepared materials is well-supported by crystalline nature.

Figure 10.7 shows XRD pattern of PA, PA sintered at 1200 °C, and PA (60%)/HCFC slag (10%)-based hybrid composite sintered at 1200 °C. Figure 10.7a shows XRD pattern of pond ash. Different peaks are observed. The highest intense peak phase is silica (ICSD analysis and indicated in Figure 10.7a). Other phases, such as mullite and alumina, are present.

Figure 10.7b displays XRD pattern of PA sintered at 1200 °C. Various peaks are observed in the pattern. The most intense peak is found at 26° (2Θ) in the XRD pattern (Figure 10.7b) and the phase is silica (ICSD analysis). Other phases such as mullite and alumina are present.

Figure 10.7c exhibits XRD pattern of PA (60%)/HCFC slag (10%)-based hybrid composite sintered at 1200 °C. The most intense peak is found at 30° (2Θ) in the XRD pattern (Figure 10.7c) and the phase is clinopyroxene phase (ICSD analysis). Other phases such as silica and mullite, forsterite, spinel, and oxides of chromium phase are present. Such phases support the improved strength, stability, and resistivity of the prepared hybrid materials.

The FTIR absorption bands of processed NALCO pond ash are shown in Figure 10.8a. Presence of different absorption stretching frequency of pond ash powder corresponds to pond ash components. The stretching frequency at 562 cm^{-1} matches with Si-O-Al band. Stretching frequencies are found at 1092 and 1623 cm^{-1}. This may be accredited to Si-O-Si asymmetric band and H-O-H bending band respectively [82–89].

The FTIR spectrum of raw HCFC slag is shown in Figure 10.8b. From Figure 10.8b, a broad stretching band can be seen around 3498 cm^{-1}. The band arises due to the stretching vibrations of O-H bonds and H bending vibrations of H-O-H of interlayer adsorbed H_2O molecule. The C=O stretching band is indicated at 1412 cm^{-1} and confirms the presence of carbonate groups.

The main reason for this being the presence of chemisorbed CO_2 in raw HCFC slag. Si-O and O-Si-O bands are also found at 1069 cm^{-1} and 561 cm^{-1},

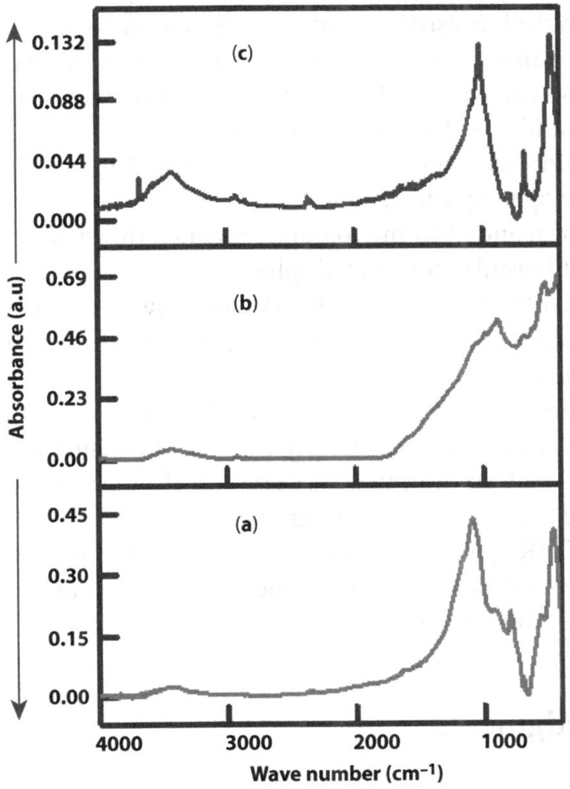

Figure 10.8 FTIR spectra of raw PA (a), raw HCFC slag (b), and PA (60%)/HCFC slag (10%)-based hybrid composite sintered at 1200 °C (c).

which indicates the presence of silicate groups. Presence of Al^{3+}-O^{2-} bonds are also observed near 808 cm^{-1}; and Fe-O stretching band is also observed at 463 cm^{-1} [81–88].

The FTIR spectrum of PA (60%)/HCFC slag (10%)-based hybrid composite sintered at 1200 °C is shown in Figure 10.8c. Peak at 1010 cm^{-1} is attributed to Si–O band in SiO_4 molecules. A weak Si–O symmetrically stretching band is observed at 656 cm^{-1}. Peaks at 1010 cm^{-1} and 870 cm^{-1} are assigned as alternating Si–O and Al–O bonds. Al–O stretching vibration is present in AlO_4 tetrahedra. There is shifting of stretching frequencies. It is observed that band of asymmetric bending of O–Si–O and O–Al–O band shifts to lower value, which confims the previous finding [81–88].

10.3 Conclusions

- PA/HCFC slag/CC/PY/k-FD/BC hybrid materials are used to develop ceramic composite by sintering them at 1200 °C.
- Electrical resistivity value of developed hybrid ceramics was estimated by two-probe technique. Maximum electrical resistance value and resistivity value were found to be 23.71 GΩ and 335×10^9 Ω.cm, respectively.
- Vickers hardness value was determined to be 393.87 for developed hybrid ceramics.
- Optical and SEM micrographs revealed the presence of distinct irregular compacted sphere.
- Microstructural and EDS analyses showed the presence of elements (O, Si, Al, Cr, Mg, etc.) in the developed hybrid materials.
- Mullite and other phases were observed in XRD pattern. TEM image analyses clearly showed crystalline phase. The SAD pattern indicated that crystals of different compound were present in the microstructure.
- In FTIR spectra, a noticeable variance was indicated. The corresponding groups of the peaks designated the formation of new bonds within the phases.

Acknowledgments

The authors want to convey their sincere thanks to GIET, University Gunupur, Rayagada, Odisha, India for providing lab facilities for their research work.

They would also like to thank CRF IIT Kharagpur for providing testing facilities. The authors would like to thank Prof. Munesh Chandra Adhikary, PG Council Chairman, Fokir Mohan University, for his invaluable guidance, advice, and constant inspiration throughout the entire program. They would also like to thank Mr. Mukteswar Mohapatra of Fokir Mohan University for his support.

References

1. https://cea.nic.in/wp-content/uploads/2020/04/flyash_201617.pdf
2. https://psr.org/wp-content/uploads/2018/05/coal-ash.pdf
3. https://www.thethirdpole.net/en/climate/coal-ash-is-a-serious-hazard-to-our-health-and-the-environment/
4. A. Bhatt, S. Priyadarshini, A. A. Mohanakrishnan, A. Abri, M. Sattlerm, and S. Techapaphawit, (2019), Case Study Physical, Chemical, and Geotechnical Properties of Coal Fly Ash: A global review, Case Studies in Construction Materials, Vol. 11, pp. e00263
5. Morrison, Garrett L. (Unity, ME), (1999), Managed composition of waste-derived fuel, patent number, 08/712312
6. https://climateconnection.org.in/updates/coal-ash-types-effects-and-solutions
7. https://en.wikipedia.org/wiki/Fly_ash
8. P.K. Rohatgi, J.K. Kim, N. Gupta, Simon Alaraj, and A. Daoud, (2006), Compressive Characteristics of A356/Fly Ash Cenosphere Composites Synthesized by Pressure Infiltration Technique Composites Part A: Applied Science and Manufacturing, Vol. 37, pp. 430-437.
9. Virendra Kumar Yadavl and M.H. Fulekar, (2018), The Current Scenario of Thermal Power Plants and Fly Ash: Production and Utilization with a Focus in India, International Journal of Advance Engineering and Research Development, Vol. 5, pp. 768-777.
10. S.U. Qadir, V. Raja, W. A. Siddiqui Mahmooduzzafar, E.F. Abd Allah, A. Hashem, P. Alam, and P. Ahmad, (2019), Fly-Ash Pollution Modulates Growth, Biochemical Attributes, Antioxidant Activity and Gene Expression in Pithecellobium Dulce (Roxb) Benth. Plants (Basel). Vol. 8, pp. 528(1-18) (doi: 10.3390/plants8120528).
11. S. Kumar, S. R. Smith, G. Fowler, C. Velis, S. J. Kumar, S, R. Arya, R. Kumar, and C. Cheeseman. (2017), Challenges and Opportunities Associated with Waste Management in India. R. Soc. open sci. Vol. 4, pp. 160764-xxx (http://dx.doi.org/10.1098/rsos.160764).
12. V. Petropavlovskaya, T. Novichenkova, M. Sulman, K. Petropavlovskii R. Fediuk, and M. Amran, Coal Ash Enrichment with Its Full Use in Various Areas

13. J. C. Egbai, (2013), Kaolin Quantification in Ukwu-nzu and Ubulu-uku using Electrical Resistivity Method, International Journal of Recent Research and Applied Studies, Vol. 14, pp. xxx-xxx.
14. O. A. Babalola and T. Akomolafe, (2008), Effects of Kaolin Particle Size and Annealing Temperature on the Resistivity of Zinc-Kaolin Composite Resistors, Journal Applied Science Environmental Management, Vol. 12, p. 113-xxx.
15. L. Li and Y. Li, (2017), Development and Trend of Ceramic Cutting Tools from the Perspective of Mechanical Processing, IOP Conf. Series: Earth and Environmental Science, Vol. 94, pp. 1-6.
16. M. Zhu, R. Ji, Z. Li, H. Wang, L. L. Liu, and Z. Zhang, (2016), Preparation of Glass Ceramic Foams for Thermal Insulation Applications from Coal Fly Ash and Waste Glass, Construction and Building Materials, Vol. 112, pp. 398-405.
17. H. E. Exner and E. Arzt, (1996), Sintering Processes, In Physical Metallurgy, Ed. R. W. Cahn and P. Haasen, 4th Ed. Elsevier Science, Amsterdam, Vol.3, pp. 2628-2662.
18. W. V. Siemens, Inventor and Entrepreneur: Recollections of Werner Von Siemens, London, England, 1966.
19. T. Tanaka and T. Imai, (2013), Advances in Nanodielectric Materials Over the Past 50 years, IEEE Electrical Insulation Magazine, Vol. 1, pp. 10-23.
20. T. Tanaka, (2005), Dielectric Nanocomposites with Insulating Properties, IEEE Transactions on Dielectrics and Electrical Insulation, Vol. 12, pp. 914-928.
21. https://www.vedantu.com/evs/uses-of-insulators
22. https://physics.stackexchange.com/questions/172076/how-does-insulating-footwear-prevent-an-electric-shock
23. https://ohsonline.com/articles/2021/09/15/six-steps.aspx
24. H. Aripin1, S. Mitsudo, B. Rahmat, S. Tani, K. Sako, Y. Fujii, K. Kikuchi, T. Saito, T. Idehara, and S. Sabchevski, (2014), Formation of Porous Clay Ceramic Using Sago Waste Ash as a Prospective Additive Material with Controllable Milling, Science of Sintering, Vol. 46, pp. 55-64 (DOI: 10.2298/SOS1401055A).
25. E. Martini1, A. Pavese2, D. Tabacchi3, D. M. Fortuna1 and A. Fortuna1, (2021), Effects of sintering temperature on microstructure and properties of sanitaryware ceramic produced with waste material Cerâmica, Vol. 67 , pp. 39-47 (http://dx.doi.org/10.1590/0366-69132021673813022)
26. S. K. Hubadillah, Z. Harun, Mohd H. D. Othman, and A. F. Ismail, and P. Gani, (2016), Effect of Kaolin Particle Size and Loading on the Characteristics of Kaolin Ceramic Support Prepared Via Phase Inversion Technique, Journal of Asian Ceramic Societies, Vol. 4, pp. xxx-xxx (DOI: 10.1016/j.jascer.2016.02.002).
27. H. Olphen, An Introduction to Clay Colloid Chemistry, John Wiley & Sons, New York, 1977.
28. T. K. Mukhopadhyay, S. Ghatak, and H. S. Maiti, (2010), Pyrophyllite as Raw Material for Ceramic Applications in the Perspective of its Pyro-Chemical Properties, Ceram. Int. Vol. 36, pp. 909–916.

29. D. Perkins, Mineralogy, Prentice-Hall Inc, Upper Saddle River, New Jersey 07458, 1998.
30. K.J. MacKenzie, S. Komphanchai, R. Vagana, (2008), Formation of inorganic polymers (geopolymers) from 2:1 layer lattice aluminosilicates, J. Eur. Ceram. Soc. Vol. 28, pp. 177–181.
31. S. Mohammadnejad, J. L. Provis, and J. S. van Deventer, (2014), Effects of Grinding on the Preg-robbing Behaviour of Pyrophyllite, Hydrometallurgy, Vol.146, pp. 154–163.
32. J. Oshitani, M. Kondo, H. Nishi, and Z. Tanaka, (2003), Separation of Silicastone and Pyrophyllite by a Gas–Solid Fluidized Bed Utilizing Slight Difference of Density, Advanced Powder Technology, Vol. 14, pp. 247–258.
33. J. Zhang, L. Hu, R. Pant, Y. Yu, Z. Wei, and G. Zhang, (2013), Effects of Interlayer Interactions on the Nanoindentation Behavior and Hardness of 2:1 Phyllosilicates, Applied Clay Science, Vol. 80, pp. 267–280.
34. D. Zhang, C. Zhou, C. Lin, D. Tong, and W. Yu, (2010), Synthesis of Clay Minerals, Applied Clay Science, Vol. 50, pp. 1–11.
35. D. Kibanova, M. Trejo, H. Destaillats, and J. Cervini-Silva, (2011), Photocatalytic Activity of Kaolinite, Catal. Commun. Vol. 12 pp. 698–702.
36. L.V. Barbosa, L. Marçal, E.J. Nassar, P.S. Calefi, M.A. Vicente, R. Trujillano, V. Rives, A. Gil, S.A. Korili, and K.J. Ciuffi, (2014), Kaolinite-Titanium Oxide Nanocomposites Prepared via Sol-Gel as Heterogeneous Photocatalysts for Dyes Degradation, Catal. Today, Vol. 246, pp. 133–142.
37. H. Wu, L. Wang, J. Zhang, Z. Shen, and J. Zhao, (2011), Catalytic Oxidation of Benzene, Toluene and P-Xylene Over Colloidal Gold Supported on Zinc Oxide Catalyst, Catal. Commun. Vol. 12, pp. 859–865.
38. H. Wu, L. Wang, Z. Shen, and J. Zhao, (2011), Catalytic Oxidation of Toluene and P-Xylene using Gold Supported on Co_3O_4 Catalyst Prepared by Colloidal Precipitation Method, J. Mol. Catal. A: Chem. Vol. 351, pp. 188–195.
39. H. Wu, G. Wu, Y. Ren, X. Li, and L. Wang, (2016), Multishelled Metal Oxide Hollow Spheres: Easy Synthesis and Formation Mechanism, Chem. - Eur. J. Vol. 22, pp. 8864–8871.
40. H. Wu, G. Wu, Y. Ren, L. Yang, L. Wang, and X. Li, (2015), Co^{2+}/Co^{3+} Ratio Dependence of Electromagnetic Wave Absorption in Hierarchical $NiCo_2O_4$–$CoNiO_2$ Hybrids, J.Mater. Chem. C Vol. 3, pp. 7677–7690.
41. H. Wu, G. Pantaleo, G. Di Carlo, S. Guo, G. Marcì, P. Concepción, A.M. Venezia, and L.F. Liotta, (2015), Co_3O_4 Particles Grown Over Nanocrystalline CeO_2: Influence Of Precipitation Agents and Calcination Temperature on the Catalytic Activity for Methane Oxidation, Catal. Sci. Technol. Vol. 5, pp. 1888–1901.
42. H. Wu, V. La Parola, G. Pantaleo, F. Puleo, A. Venezia, and L. Liotta, (2013), Ni-based Catalysts for Low Temperature Methane Steam Reforming: Recent Results on Ni-Au and Comparison with other Bi-Metallic Systems, Catalysts, Vol. 3, pp. 563–583.
43. G. Wu, Y. Cheng, Y. Ren, Y. Wang, Z. Wang, and H. Wu, (2015), Synthesis and Characterization of C-Fe_2O_3@C Nanorod-Carbon Sphere Composite

And Its Application As Microwave Absorbing Material, J. Alloy. Compound, Vol. 652, pp. 346–350.
44. H. Wu and L. Wang, (2011), Shape Effect of Microstructured CeO_2 with Various Morphologies on CO Catalytic Oxidation, Catal. Commun. Vol. 12, pp. 1374–1379.
45. H. Wu and L. Wang, (2014), Phase Transformation-Induced Crystal Plane Effect of Iron Oxide Micropine Dendrites on Gaseous Toluene Photocatalytic Oxidation, Applied Surface Science, Vol. 288, pp. 398–404.
46. L. Chen, C.H. Zhou, S. Fiore, D.S. Tong, H. Zhang, C.S. Li, S.F. Ji, and W.H. Yu, (2016), Functional Magnetic Nanoparticle/Clay Mineral Nanocomposites: Preparation, Magnetism and Versatile Applications, Applied Clay Science, Vol. 127–128, pp. 143–163.
47. E.M.S. Azzam, S.M. Solyman, and A.A. Abd-Elaal, (2016), Fabrication of chitosan/Ag Nanoparticles/Clay Nanocomposites for Catalytic Control on Oxidative Polymerization of Aniline, Colloids Surface A, Vol. 510, pp. 221–230.
48. A. Raju, V. Lakshmi, R.K.V. Prataap, V.G. Resmi, T.P.D. Rajan, C. Pavithran, V.S.Prasad, and S. Mohan, (2016), Adduct Modified Nano-Clay Mineral Dispersed Polystyrene Nanocomposites as Advanced Corrosion Resistance Coatings For Aluminum Alloys, Applied Clay Science, Vol 126, pp. 81–88.
49. L. Xia, H. Zhong, G. Liu, Z. Huang, Q. Chang, and X. Li, (2009), Comparative Studies on Flotation of Illite, Pyrophyllite and Kaolinite with Gemini and Conventional Cationic Surfactants, Trans. Nonferr. Met. Soc. China, Vol. 19, Vol. 446–453.
50. H. Zhong, G. Liu, L. Xia, Y. Lu, Y. Hu, S. Zhao, and X. Yu, (2008), Flotation Separation of Diaspore from Kaolinite, Pyrophyllite and Illite using Three Cationic Collectors, Miner. Eng. Vol. 21, pp. 1055–1061.
51. L. Xia, H. Zhong, and G. Liu, (2010), Flotation Techniques for Separation of Diaspore from Bauxite using Gemini Collector and Starch Depressant, Trans. Nonferr. Met. Soc.China Vol. 20, pp. 495–501.
52. Y. Hu, X. Liu, and Z. Xu, (2003), Role of Crystal Structure in Flotation Separation of Diaspore from Kaolinite, Pyrophyllite and Illite, Miner. Eng. Vol. 16, pp. 219–227.
53. M. Erdemoğlu and M. Sarıkaya, (2002), The Effect of Grinding on Pyrophyllite Flotation, Miner. Eng. Vol. 15, pp. 723–725.
54. Y. Hu and X. Liu, (2003), Chemical Composition and Surface Property of Kaolins, Miner. Eng. Vol. 16, pp. 1279–1284.
55. S.K. Jena, S. Singh, D.S. Rao, N. Dhawan, P.K. Misra, and B. Das, (2015), Characterization and Removal of Iron from Pyrophyllite Ore for Industrial Applications, Miner. Metall. Process. Vol. 32, pp. 102–110.
56. F.K. Crundwell, On the mechanism of the flotation of oxides and silicates, Miner. Eng. 95 (2016) 185–196.
57. W. Stumm, Chemistry of the solid-water interface: processes at the mineralwater and particle-water interface in natural systems, Soil Sci. 156 (1992) 21–72.

58. Liu L, Gorsey AE (2007) High-pressure phase transitions of the feldspars, and further characterization of lingunite. Int Geol Rev 49:854–860.
59. (a). K. Noritomi (1955) Studies on the Change of Electrical Conductivity with Temperature of a Few Silicate Minerals, Science ReportsTohoku Univ Ser, Vol. 5, pp. 119–126.
(b). N. Khitarov, and A. Slutskiy (1965) the Effect of Pressure on the Melting Temperatures of Albite and Basalt (based on Electroconductivity Measurements), Geochem Int Vol. 2, pp. 1034–1042.
(c). H. Mizutani, and H. Kanamori, (1967) Electrical Conductivities of Rockforming Minerals at High Temperatures, J Phys Earth, Vol. 15, pp. 25–31.
(d). A. J. Piwinskii, and A. Duba (1974) High Temperature Electrical Conductivity of Albite, Geophys Res Lett Vol. 1, pp. 209–211.
(e). A. J. Piwinskii, A. Duba, and P. Ho (1977), The Electrical Conductivity of Low and High Albite Throughout its Melting Interval at 100 kPa, Can Mineral, Vol. 15, pp.196–197.
60. (a) N. S. Bagdassarov, and N. Delépine (2004), A–B Inversion in Quartz from Low Frequency Electrical Impedance Spectroscopy, J Phys Chem Solids, Vol. 65, pp. 1517–1526.
(b) N.S. Bagdassarov, J. Maumus, B. Poe, and V. Bulatov (2004), Pressure Dependence of Tg in Silicate Glasses from Electrical Impedance Measurements, Phys Chem Glasses B, Vol. 45, pp. 197–214.
61. (a) X. Yang and F. Heidelbach (2012), Grain Size Effect on the Electrical Conductivity of Clinopyroxene. Contrib Mineral Petrol, Vol. 163, pp. 939–947.
(b). X. Yang, H. Keppler, C. McCammon, and H. Ni, (2012), Electrical Conductivity of Orthopyroxene and Plagioclase in the Lower Crust, Contrib Mineral Petrol, Vol. 63, pp. 33–48.
62. H. Hu, H. Li, L. Dai, S. Shan, and C. Zhu (2011), Electrical Conductivity of Albite at High Temperatures and High Pressures, Am Mineral, Vol. 96, pp. 1821–1827.
63. I. R. Wilson, (1998), The Constitution, Evaluation and Ceramic Properties of Ball Clays. Cerâmica, Vol. 44 (287–288), pp. 88–117.
64. (a) M. Dondi, G. Guarini, and M. Raimondo, and F. Salucci, (2003), Influence of Mineralogy and Particle Size on the Technological Properties of Ball Clays for Porcelain Stoneware Tiles, Tile Brick International Vol. 20, pp. 2–11.
(b) M. Dondi, C. Iglesias, E. Dominguez, G. Guarini, and M. Raimondo, (2008), The Effect of Kaolin Properties on their Bahaviour in Ceramic Processing as Illustrated by a Range of Kaolins from the Santa Cruz and Chubut Provinces, Patagonia (Argentina), Applied Clay Science, Vol. 40, pp. 143–158.
(c). L. Stoch, (1964), Wpływ Składu Mineralnego Na Niektóre Własciwości Technologiczne Glin Kaolinitowych, Ceramika NR 2 (in Polish, with English abstract).

65. (a) V. Das Kshama, B. V. Mohan, M. Lalithambika, and C. G. R. Nair, (1992), Sintering Studies on Plastic Clays. Ceramics International, Vol. 18, pp. 359–364. (b). K. Galos, (2007). Ball clays in central and Eastern Europe, Industrial Minerals, Vol. 2, pp. 36–43.
66. Musa Alhassan, Mustapha Mohammed, and E. Mesaiyete, (2014), Influence of Clay Mineralogy on Plasticity of Lateritic Soils, IOSR Journal of Mechanical and Civil Engineering, Vol. 11, pp. 18-21 (DOI: 10.9790/1684-11341821).
67. F. A. de Andrade, H. Ali Al-Qureshi, and D, Hotza, (2011), Measuring the plasticity of clays: A review, Applied Clay Science, Vol. 51, pp. 1-7 (DOI: 10.1016/j.clay.2010.10.028).
68. Y. S. Liu, L. L. Ma, Y. S. Liu, and G. X. Kong, (2006), Investigation of Novel Incineration Technology for Hospital Waste, Environmental Science and Technology, Vol. 40, pp. 6411-xxx.
69. Y. S. Liu, (2005), A Novel Incineration Technology Integrated with Drying, Pyrolysis, Gasification and Combustion of MSW and Ashes Vitrification, Environmental Science and Technology, Vol. 39, pp. 3855-xxx.
70. S. B. Kondawar, A. D. Dahegaonkar, V. A. Tabhane, and D. V. Nandanwar, (2014), Thermal and Frequency Dependence Dielectric Properties of Conducting Polymer/Fly Ash Composites, Advanced Materials Letters, Vol. 5, pp. 360-xxx.
71. A. Pattnaik, S. K. Bhuyan, S. K. Samal, A. Behera, and S. C. Mishra, (2016), Dielectric Properties of Epoxy Resin Fly Ash Composite, IOP Conference Series: Materials Science and Engineering, Vol. 115, pp. 1-xxx.
72. M. Mishra, A. P. Singh, and S. K. Dhawan, (2013), Utilization of Fly Ash-A Waste By-product of Coal for Shielding Application, Journal of Environmental Nanotechnology, Vol. 2, pp. 74-xxx.
73. J. P. Dhal, and S. C. Mishra, (2013), Investigation of Dielectric Properties of a Novel Hybrid Polymer Composite using Industrial and Bio-waste, Journal of Polymer Composites, Vol. 1, pp. 22-xxx.
74. K. Singh, T. Quazia, S. Upadhyaya, and P. Sakharkarb, (2005), Development of Low Permittivity Material using Fly Ash, Indian Journal of Engineering & Materials Sciences, Vol. 12, pp. 345.
75. E. P. A. Chinese, (1996), China Identification Standard for Hazardous Waste Toxicity (GB5085.3-1996), Chinese Environmental Science Press, Beijing.
76. M. Panigrahi, R. I. Ganguly, R. R. Dash, (2018), Development of Novel Constructional Material from Industrial Solid Waste as Geopolymer for Future Engineers IOP Conference Series: Materials Science and Engineering, Vol. 410, pp. 1-12.
77. M. Panigrahi, R. I. Ganguly, R. R. Dash, (2018), An Advanced Cured High Carbon Ferrochrome Slag (HCFCS) Geopolymer (GP): A Constructional Materials IOP Conference Series: Materials Science and Engineering, Vol. 410, pp. 1-15.
78. M.K. Panigrahi, (2021), Investigation of Structural, Morphological, Resistivity of Novel Electrical Insulator: Industrial Wastes, Bull. Sci. Res. Vol. 3, pp. 51-58.

79. X. Liu and M. Bai, (2017) Effect of chemical composition on the surface charge property and flotation behavior of pyrophyllite particles, Advanced Powder Technology, Vol 28, pp. 836–841.
80. X. Liu and M. Bai, (2011), Composition and Ceramic Properties of Ball Clays for Porcelain Stoneware Tiles Manufacture in Poland, Applied Clay Science, Vol. 51, pp. 74–85.
81. N. F. Ayoob, J. M. Juoi, Z. M. Rosil, and N R. Rosil, (2011), Characterization and Properties of Sintered Glass-Ceramic Produced from Recycling of Glass by using Pressureless Method, Key Engineering Materials, Vol. 471-472 pp 933-938 (doi:10.4028/www.scientific.net/KEM.471-472.933).
82. B.J. Saikia, G.K Rao, and P. Sarathy, (2010), Fourier Transform Infrared Spectroscopic Characterization of Kaolinite from Assam and Meghalaya, North Eastern India, International Journal of Modern Physcs, Vol. 1, pp. 206-xxx.
83. R. P. dos Santos, J. Martins, C. Gadelha, B. Cavada, A.V. Albertini, F. Arruda, M. Vasconcelos, E. Teixeira, F. Alves, J.L. Filho, and V. Freire, (2014), Coal Fly Ash Ceramics: Preparation, Characterization, and Use in the Hydrolysis of Sucrose. Scientific World Journal, Vol. 2014, pp. 01.
84. A. Jose, M. R. Nivitha, M. Krishnan, and R. G. Robinson, (2020), Characterization of Cement Stabilized Pond Ash using FTIR Spectroscopy, Construction and Building Materials, Vol. 263, pp.120136-xxx (DOI: 10.1016/j.conbuildmat.2020.120136).
85. Y. Liu, F. Zeng, B. Sun, P. Jia, and I. T. Graham, (2019), Structural Characterizations of Aluminosilicates in Two Types of Fly Ash Samples from Shanxi Province, North China, Minerals, Vol. 9, pp. 358-16 (doi:10.3390/min9060358).
86. P. Rożek, M. Król, and W. Mozgawa, (2018), Spectroscopic Studies of Fly Ash-Based Geopolymers, Spectrochim Acta - Part A Molecule, Biomolecule Spectroscopy Vol. 198, pp. 283-289 (https://doi.org/10.1016/j.saa.2018.03.034).
87. S. Kumar, F. Kristály, and G. Mucsi, (2015), Geopolymerisation Behaviour of Size Fractioned Fly Ash, Advanced Powder Technology, Vol. 26, pp. 24-30 (https://doi.org/10.1016/j.apt.2014.09.001).
88. Z. Kledyński, A. Machowska, B. Pacewska and I. Wilińska, (2017), Investigation of Hydration Products of Fly Ash-Slag Pastes, Journal of Thermal Analysis and Calorimetry, Vol. 130, pp. 351-363 (https://doi.org/10.1007/s10973-017-6233-4).
89. D. Bondar and R. Vinai, (2022), Chemical and Microstructural Properties of Fly Ash and Fly Ash/Slag Activated by Waste Glass-Derived Sodium Silicate, Crystals, Vol. 12, pp. 913-12 (https:// doi.org/10.3390/cryst12070913).

11

Applications, Challenges and Opportunities of Industrial Waste Resources Ceramics

Muktikanta Panigrahi[1]*, Ratan Indu Ganguly[2] and Radha Raman Dash[3]

[1]Department of Materials Science, Maharaja Sriram Chandra Bhanja Deo University, Keonjhar Campus, Odisha, India
[2]Department of Metallurgical Engineering, National Institute of Technology, Rourkela, Odisha, India
[3]CSIR-National Metallurgical Laboratory, Jamshedpur, Jharkhand, India

Abstract

Wastes from the ferrochrome industry pose many health problems. Recycling and efficient utilization of the voluminous slag generated by this industry is a great challenge. Conversion of these materials into marketable value-added products will help to solve this problem to a greater extent.

The main objective of this work is to review the available literature and R&D status in this direction. A detailed analysis of raw materials (i.e., FeCr slag) is discussed and additional scope of work in this field is discussed.

The presence of crystalline phases imparts distinctive properties such as strength, stability and refractoriness. Results from the present work suggest that raw slag material has potential to develop ceramic, refractory, and civil construction products. Chemical composition of ferrochromium slag is suited for manufacturing high-temperature refractory ceramics from it. Utilization of ferrochromium slag in the construction sector, i.e., building material and road construction, will share the burden of natural resources.

Keywords: FeCr slag, environmental issues, ceramic products, applications, opportunities, challenges

**Corresponding author*: muktikanta2@gmail.com

Muktikanta Panigrahi, Ratan Indu Ganguly and Radha Raman Dash (eds.) High Electrical Resistant Materials: Ferrochrome Slag Resource Ceramics, (241–256) © 2024 Scrivener Publishing LLC

11.1 Introduction

Stainless steel industries consume the largest quantity of ferrochrome. High-carbon ferrochrome is produced by carbothermic reduction. Production of ferrochrome slag is ~1.5 tons. Global ferrochromium slag production is ~12–16 million tons per year [1]. Slag composition is an important parameter in ferrochrome smelting technology. The majority of ferrochromium slag is discarded and sent to landfills, which has created a huge waste deposit over the years.

Ferrochrome slag is an inorganic waste material. However, it may be used for different work to help keep Earth free from contamination [2]. Many researchers are addressing the problem of chromium (i.e., Cr^{+3} or Cr^{+6}). There is a possibility of slow release of Cr^{+6} into the environment in the long run. Therefore, profitable use of these wastes will safeguard the environment. This is the challenge for scientists and technologists.

11.2 Different Ways of Utilizing FeCr Slag Waste

Presently, researchers and technologists are putting effort into seeking ways to utilize FeCr slag for development of ceramics. These slags are fashioned with different value-added products such as ceramics products, porous insulation refractory, dense refractories, ceramic tiles, waste glass powders, glass, glass-ceramics, mullite, etc. The products developed will help in many engineering applications. Many companies worldwide have been using slags for constructing roads, buildings and landfills. The applications described in this section include i) ceramic and refractory application, ii) road and building construction application, iii) hazardous hexavalent Cr treatment, iv) manufacture of cost-effective and high-strength MMCs, v) valuable metals recovery from slag, vi) rip rap application, and vii) stone column application.

11.2.1 Ferrochromium Slag for Ceramic and Refractory Application

FeCr slag contains high melting phases such as spinel, forsterite, etc., and low melting phases such as enstatite and siliceous glass enriched with oxides of calcium, aluminum, chromium and iron. Because of low melting phases, melting temperature of slag decreases. Pseudo ternary phase (MgOAl2O3-SiO2-CrO) having 10% CrO is used to prepare refractory. Refractory behavior can be improved by increasing concentrations of

MgO, Al2O3 and Cr2O3 separately or together. There are several ways to prepare ferrochromium slag-based refractory materials. Figure 11.1 shows optical images of concrete/cement cinder blocks texture/pattern (A) and HCFC slag-based refractories (B).

Addition of desired additives to the molten slag helps in production of refractories production. This will save energy. In this process, molten ferrochromium slag is preheated in a special slag melting furnace. MgO or Cr2O3 is added to it. Materials obtained by this method are used as fused grain to prepare refractories [3].

Ashimov *et al.* [4][a,b] studied high-carbon ferrochromium slag (Aktyubinsk ferroalloy factory). Ferrochromium slag contains high amount of refractory phases (i.e., forsterite, magnesia spinel, chromite, chrome spinel or picotite). Occurrence of low melting phases (such as enstatite and schlorlomite) reduces its refractoriness. Due to the presence of glassy phase and ferrochrome metal in the slag, FeCr slag shows reduced refractoriness. The presence of glassy phases is due to calcium-aluminosilicate, melilite or enstatite. Enhancement of refractoriness of high-carbon ferrochromium slag is possible by changing low melting phases into high refractory phases. Addition of MgO overcomes the injurious effect of FeCr slag. By 20% MgO addition to the slag, refractoriness is improved to 1740 °C.

Kumar *et al.* [5] made castable refractories from ferrochromium slag. They observed better spalling resistance of their refractories in comparison to conventional castable products. High-carbon ferrochromium slag, calcined bauxite, high alumina cement and small amount of microsilica mixture are being used for production of castable refractories. Castable products are capable of withstanding 1300 °C without being distorted. Superior mechanical strength, thermal shock resistance and increase in density of the product make them qualified for use in higher temperature environment. The enhancement in properties is attributed to formation of mullite phase at high-temperature sintering where they are made.

Heat resistant, fire resistant and alkali resistant mineral wool can be manufactured from ferrochromium slag by changing its viscosity. Aluminum oxide/silicon oxide is added into molten slag to adjust deliberation temperature of slag. Low-temperature glass wools can be manufactured when silicon oxide and aluminum oxide maintained at a ratio of 2.5-5.0:1 are added to ferrochromium slag. While high-temperature application glass wools can be manufactured if aluminium oxide is added in the range of 20–30% by weight.

Air-cooled ferrochromium slag is acid resistant and superior to granite in terms of strength properties. Guloyan *et al.* [6] studied the utilization of ferrochromium slag for production of bottles. They found that

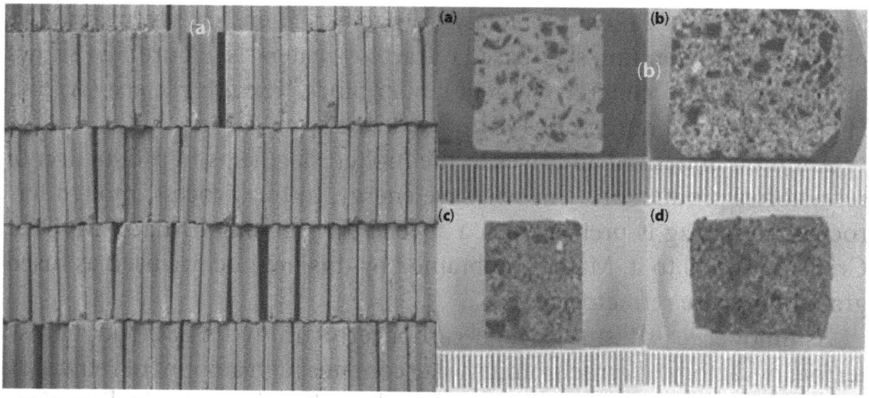

Figure 11.1 Optical images of concrete/cement cinder blocks texture/pattern (a) and HCFC slag-based refractories (b).

ferrochromium slag, which acts as a coloring agent and lowers melting temperature, can be used for making bottles. Chromic oxide addition improves melting temperature and working properties of container glass in comparison to iron oxide and manganese oxide. Thus, addition of ferrochromium slag paved the way for production of low-cost glass material. Ferrochromium slag does not contain free chromic oxide and all components of the ferrochromium slag are easily soluble in glass melts. This favors homogeneous coloring of glass during glass manufacturing process compared to other coloring additives [12].

11.2.2 Ferrochrome Slag in Road/Building Construction Applications

Due to the low leachability of chromium, FeCr slag is used for road construction. A significant amount of Cr and other heavy metals are absorbed by plants [7]. Chromium-bearing slag mixed with blast furnace slag cement stabilizes chromium. Therefore, concrete aggregate made from this mixture will be safer [7, 8]. That is why ferrochromium slag is more compatible with the environment for concrete products having excellent compressive strength. This is because of the occurrence of sulfide compounds and ferrous salts in blast furnace slag. These two compounds are responsible for reduction of Cr(VI) to Cr(III). Chromium leaching from concrete made with FeCr slag aggregate and blast furnace slag-based cement is insignificant. This is well-supported by the Indian regulatory discharge standard and U.S. Environmental Protection Agency standard [9].

11.2.2.1 *Ferrochrome Slag Concrete*

Ferrochromium slag-based concrete pavement is more economical than Portland cement-based concrete pavement. Air-cooled ferrochromium slag is used in high-strength concrete as coarse aggregate while water-cooled granulated ferrochromium slag is suitably used as fine aggregate in concrete work. Zelic [10] reported that air-cooled slag is more suitable for concrete structure. Air-cooled ferrochromium slag concrete pavement has better compressive strength, wear resistance and specific weight if compared with commercial Portland cement-based pavement. Figure 11.2 shows optical photographs of HCFC slag-based concrete and cement slag stock.

Utilization of ferrochromium slag with fly ash is one of the effective ways to use ferrochromium slag as construction material. Gencel *et al.* [11] studied the combined effects of fly ash and ferrochromium slag on concrete performance. In their study, cement is replaced by fly ash and coarse lime stone aggregate by coarse ferrochromium slag aggregate in fresh concrete in appropriate proportion. Fly ash lowers compressive strength, wear resistance, tensile strength and splitting tensile strength but enhances freeze-thaw durability in concrete structure. Moreover, ferrochromium slag increases strength and abrasive wear resistance of concrete. Fly ash and ferrochromium slag-based concrete enhances the overall properties.

11.2.3 Hazardous Hexavalent Cr Treatment

Erdem *et al.* [12] successfully eliminated hexavalent chromium from ferrochromium slag by jigging it in a water stream. During the jigging operation, ferrochromium slag and sulfuric acid in optimum quantity are used

Figure 11.2 Optical photograph of HCFC slag-based concrete (a) and cement slag stock (b).

in a closed system. Reduction method is adopted to remove hexavalent chromium. In the reduction process, hexavalent Cr becomes trivalent Cr. Trivalent Cr is precipitated out.

11.2.4 Manufacture of Cost-Effective High-Strength MMCs

Aluminum-based metal matrix composites (MMCs) exhibit attractive properties (i.e., high thermal conductivity, low density, high specific strength to light weight, stiffness, dimensional stability) at elevated temperatures [13][a-f]. 7075 Aluminium alloy is a good abrasive resistant, wear resistant and heat-treatable alloy. It is used in a wide range of applications such as automobile, biomedical, aerospace industries, general engineering sectors, and particularly in aerospace and structural components [14]. In addition, Al-based metal matrix composites (MMCs) are used for tribological applications. Among the various reinforcements, FeCr slag particles are one of the low-priced reinforcements existing in large quantities. Harder reinforcement particles improve mechanical properties (like hardness) and tribological properties (like wear resistance). Particle size influences the composite properties. Size of particle varies from micrometer to nanometer. It shows a dramatic improvement in strength. In order to prevent agglomeration and particle clustering, reinforced particle size should be homogeneous. Composite materials are manufactured by different methods, among which the stir casting technique is popularly used for preparing Al-MMCs, which ensures uniform distribution of particles in the matrix.

11.2.5 Valuable Metals Recovery from FeCr Slag

For production of ferrous and non-ferrous metals by pyrometallurgical process, a huge amount of slags are generated from different industries. Slag contains a mixture of metal oxides/sulfides and silicates with glassy appearance. Depending on the slag source, slags contain heavy and hazardous metals. The dumping of slags consumes a large area of cultivable land and causes water and soil pollution. Due to possible leaching of heavy metals, water and soil becomes contaminated with harmful metals and threatens human life by causing different types of diseases (e.g., anemia, stomach cancer, etc.). More intake of iron from contaminated water may cause hemochromatosis [15]. Overexposure to dry or wet cement slag causes irreparable eye, skin or respiratory disorders. Since the hexavalent chromium in FeCr slag is carcinogenic, it may cause skin, gastrointestinal and lung cancers in humans and animals. So, it is of utmost importance to recover chromium metal from FeCr slag in order to protect our environment. Furthermore, slag is a rich source of metal that can be recovered by adopting different methods. This is also technologically viable. In the current scenario, various strategies are in practice to recover heavy metal [16–23].

11.2.5.1 *Recovery of Iron*

Iron can be recovered from slag by adopting feasible processing technique(s) such as crushing, grinding, gravimetric and magnetic separation/ leaching, and pressure leaching/pyrometallurgical/ pyrohydrometallurgical treatments [18]. A few reports are available on recovery of iron from slag, which are discussed below. In all types of slags, iron is present as iron oxide (FeO) or metallic iron.

Shen and Forssberg [24] recovered iron from basic oxygen furnace (BOF), electric arc furnace (EAF), and ladle furnace (LF) slag using the magnetic separation method (various magnetic machines and screens). Recovery of iron has been reported to be 8%.

Mazurek [25] reported the recovery of iron from post-leaching residues. Iron was also recovered by ion exchange method using three resins. The extent of recovery is reported to be 91.4%, 90.7% and 92.4% iron.

11.2.5.2 *Recovery of Chromium*

Chromium (Cr) is present in stainless steel slag, charge chrome slag and vanadium slag (as Cr_2O_3). However, there is a possibility to convert hazardous CrO_3 after disposal or during processing. As chromium is a toxic metal, different methods are employed to recover it from slags. Methods of Cr recovery are crushing, magnetic separation, gravity separation, smelting reduction and leaching technique [24, 26]. In stainless steel slag, Cr_2O_3 is most stable among various Cr oxides at 298 K in air. At higher temperatures, Cr_2O_3 is easily oxidized to CrO_3 or Cr^{6+}. Different methods have been proposed to avoid oxidation. It is proposed that aluminum is the best reductant for Cr oxide. However, carbon and silicon are other reductants. Carbon and silicon are widely used for reduction of a large amount of Cr oxide in the slag [27].

Das *et al.* [28] reported the recovery of Cr from charge chrome slag by a two-step magnetic separation process. In this process, 57.1% of chromium is recovered in the form of Cr_2O_3.

Choudhury *et al.* [29] revealed 11.5% Cr containing charge chrome slag. Charge chrome slag produces 55% Cr having particle sizes 5 mm.

Holappa and Xiao [30] have reported 90–99% of Cr recovery from charge chrome slag by leaching process.

Li *et al.* [31] used stepwise roasting method to recover Cr from high Cr content vanadium slag. In this method, 96.4% Cr is recovered from leaching residue. This is an environment friendly process for optimal conditions.

Kononova *et al.* [32] recovered Cr(VI) from aqueous solution by adsorption onto long-chained crosslinking agents (LCA). In the sorption of hexavalent Cr, anion exchangers, i.e., AN-108 (synthesized from copolymers of methylacrylate aminated with ethylenediamine), are used. Due to their high selectivity and excellent kinetic properties, these sorbents can recover 100% Cr.

11.2.5.3 Recovery of Mg

Magnesium (Mg) is a crucial metal for high technological applications such as aerospace, military, etc. It is the lightest known structural metal which is cost-effective [33]. Also, Mg has found wide applications in agriculture, particularly in the synthesis of fertilizers. It is a secondary nutrient crucial for chlorophyll synthesis [34].

Currently, population growth and technological advancement are both occurring. Therefore, there is a demand for increasing production of Mg. Since the primary sources of Mg are magnesite ($MgCO_3$), dolomite $CaMg(CO_3)_2$, and seawater, it is imperative to exploit those waste sources which are rich in Mg in order to meet growing global demand [35, 36].

Bradford-Hartke *et al.* [37] have reported that FeCr slag has potential use in different fields such as construction, glass ceramics, and recovery of valuable metals. Extraction methods used to recover metals such as Mg from slag materials are cheap compared to technologies adopted to recover Mg from primary sources (Pidgeon process and thermal reduction). For recovering Mg, the leaching process is adopted. Using leachate in its impure state is envisaged for processes, such as struvite precipitation, for socioeconomic gain. Struvite ($NH_4MgPO_4.H_2O$) is a slow-release mineral fertilizer which has commercial value.

Shaddel *et al.* [38] estimated Mg content in seawater, which is 1300 mg/L. They used the struvite precipitation method. Despite other cations, such as calcium, the process tends to result in undesirable side reactions.

11.2.5.4 Recovery of Aluminum, Sodium and Potassium

Aluminum is present in slags in the form of oxide (Al_2O_3), while sodium and potassium are present in salts. Metallic aluminium can be recovered from slag and incineration slag by mechanical method and leaching followed by filtration and evaporation. Conventional treatment of salt slag consists of grinding, sieving for separation of big lumps followed by water leaching at ambient or higher temperature and pressure to dissolve salt in water, and then filtration and evaporation to recover metals.

Reynolds and Olper [39] proposed a five-step process for recovery of Al and other metals from salt slag. Metallic Al is separated from other salts and oxides by sequential crushing and screening due to difference in physical properties. From this separation method, 7% yield of Al is obtained from salt slag. After separation of Al, salts and oxides are further subjected to water leaching. In this process, salts get dissolved in water, while oxides are insoluble. Such mixture of liquid and solid is filtered by thickener filter material. It is washed to produce oxides from underflow and overflow rich in salts and is passed through evaporation and crystallization steps. During the steps, some toxic gases are produced. Ammonia gas is removed by passing through a scrubber and a column of sulfuric acid. Then, residual gases are removed by passing through a multi-stage absorber or are burnt to avoid toxic effects. The salt solution after filtration is passed through an evaporator to produce final crystals of NaCl and KCl with approximately 50% yield. From the resulting slurry, the remaining Al are washed, separated by a rotating screen, dried and 80% Al recovered. Al can also be obtained from bottom ash of incineration slag by mechanical separation incorporating drying, crushing, sieving, magnetic separation and eddy current separation method.

11.2.6 Rip Rap Application

Because of their large boulder size with weight, water absorption property, and high compressive strength in soaked condition, FeCr slag is required for Rip Rap applications. It stabilizes the slopes of earthen embankments, particularly highway embankments, reservoir bunds and canal embankments. Hence, it is suitable for aspects of slope stability [40, 41]. Figure 11.3a and 11.3b indicate Rip Rap application.

11.2.7 Stone Column Application

Ferrochrome (FeCr) slag particles have excellent mechanical properties such as strength, toughness, etc. The particle size of FeCr slag ranges from 6.0 mm to 70 mm. Shear parameters of FeCr slag determine frictional characteristics. So, FeCr slag has advantages for use in stone columns as a construction material to support structures (such as embankments and oil

Figure 11.3 Photographs of Rip Rap erosion stock (a&b) [42, 43].

Figure 11.4 Photographs showing slag stones (a&b) [46, 47].

storage tanks in soft clays). Figure 11.4a and 11.4b shows slag stones. These are used in stone column application [44, 45].

11.3 Challenges

The world is experiencing environmental pollution due to generation of industrial wastes, by-products, gases, etc. Wastes produced from thermal power plants and ferrochrome industries cause global warming. Therefore, scientists, engineers and researchers are facing many challenges to solve these problems. One way to tackle pollution is to control the generation of industrial wastes. Another way is power generation by renewable sources (such as solar, nuclear, wind, etc.). Different ways of achieving a carbon-free atmosphere are being envisaged by materials scientists. They are looking at inventing new materials at lower cost in order to create a carbon-free world that is more compatible with human life.

Cement continues to be a very important construction material. However, cement is produced by a high-energy process, which causes an increase in global temperature and global pollution. If cement can be replaced by other materials, then pollution can be controlled to a greater extent.

The present work has attempted to create new materials, i.e., high resistance ceramics, for different applications in the ceramic sectors. As stated earlier, many scientists are working on ceramic production using fly ash, pond ash, and ferrochrome slag waste resource materials.

11.4 Opportunity

Replacement of porcelain-based ceramic using industrial waste will be a boon in the society. Another important aspect of its use is the economic factor. Encouraging entrepreneurial activities will help to solve employment problems. In addition, industrial waste-based ceramic will immensely benefit the modern world. While thinking of such materials, one must consider ease of production and availability of cheaper raw materials. All of these factors will provide an opportunity for entrepreneurs starting a business. It will also help countries development at a faster pace.

In conclusion, the following points are drawn from the above discussion:

- Industrial waste-based ceramic is a cost-effective material since raw materials are available as waste.
- It utilizes waste materials, which will protect the Earth.
- It simplifies the production process, where high temperature is needed.
- It makes production easier.
- It replaces costlier conventional ceramics.
- It solves employment problems by increasing entrepreneurial activities.

All of these factors will help to build a modern nation.

11.5 Conclusions

In the 21st century, there continues to be a greater societal demand to protect our environment with a planning strategy of sustainable development.

Sustainable development of waste products is one of the routes to reduce environmental pollution. Recycling of ceramic waste is also a useful technique to protect our environment. Looking towards high-speed development, availability of raw materials, and non-renewable resources, waste is used fully or partially to avoid critical situations. Hence, sustainable development helps to improve socioeconomic conditions. Reviews have been made of the comprehensive studies on utilization of FeCr slag. The different ways in which FeCr wastes are used as the following:

- Nanomaterials (i.e., Fe_3O_4, MFe_2O_4 (M- Mn, Cu, Zn, Ag), magnetic biochar, $MgCr_2O_4$, Cr_2O_3, CuO, copper ferrite, etc.) are synthesized from FeCr by physico-chemical methods.
- FeCr slag makes production of building ceramics possible.
- Binary or ternary combinations of ingredients with FeCr slags are permitted to prepare glass-ceramics such as floor and wall tiles, bench tops, sewer pipes and many others with valuable properties.

Acknowledgments

The authors would like to thank Prof. Munesh Chandra Adhikary, PG Council Chairman, Fokir Mohan University, for his invaluable guidance, advice, and constant inspiration throughout the entire process. They would also like to thank Mr. Mukteswar Mohapatra of Fokir Mohan University for his valuable support.

References

1. P. Richard, (2015), Overview of the Global Chrome Market, INDINOX Stainless Steel Conference-1. January 25-26, Ahmedabad.
2. Authors name (2016), A Short Review on Utilization of Ferrochromium Slag, Mineral Processing and Extractive Metallurgy Review-An International Journal, Vol. xxx, pp.1-42 (DOI: 10.1080/08827508.2016.1168415).
3. G.I. Zaldat, V.V. Kamyshnikov, S.A. Suvorov, V.S. Volkov, G.A. Koshkin, M.G. Yurman, Y.I. Ostrovskii and V.D., Naryzhnyi, (1984), Fused Refractory Materials Based on Magnesia Slags. Refractory, Vol. 25, pp. 125-130.
4. (a) U.B. Ashimov, A.V. Bolotov, and D.S. Satvaldiev, (1984), Using Ferrochromium Slags in the Preparation of Forsterite Refractory, Refractory, Vol. 25, pp. 210-213.

(b) V.A. Perepelitsyn, Y.I. Savchenko, V.M. Bezhaev, and S.N. Tabatchikova, (1986), Petrochemical Calculation in the Production of Forsterite-Spinel Refractories fFrom Carbon Ferrochromium Slags, Refractory, Vol. 27, pp. 568-572.

5. P.H. Kumar A. Srivastava, V. Kumar, M.R. Majhi, and V.K. Singh (2014), Implementation of Industrial Waste Ferrochromium Slag in Conventional and Low Cement Castables: Effect of Calcined Alumina, Journal of Asian. Ceramic Society, Vol. 32, pp. 371–379.
6. Y.A. Guloyan, K.S. Katkova, V.F. Smirnov, A. F. Sudin, M. A. Basova and S. G. Korobov 1972), Use of Ferrochromium Slags in the Production of Bottles. Glass and. Ceramics, Vol. 29, pp. 527-529.
7. B.B. Lind, A.M. Fallman, and L.B. Larsson, (2001), Environmental Impact of Ferrochromium Slag in Road Construction, Waste Management, Vol.21, pp. 255-264.
8. A. Kindness, A. Macias, and F.P. Glasser, (1994), Immobilization of Chromium in Cement Matrices, Waste Management, Vol. 14, pp. 3-11.
9. C.R. Panda, K.K. Mishra, K.C. Panda, B.D. Nayak, B.B. Nayak, (2013), Environmental and Technical Assessment of Ferrochromium Slag as Concrete Aggregate Material, Construction and Building Materials, Vol. 49, pp. 262-271.
10. J. Zelic (2005), Properties of Concrete Pavement Prepared With Ferrochromium Slag as Concrete Aggregate, Cement and Concrete Research, Vol. 35, pp. 2340-2349.
11. Gencel O., Koksal F., Ozel C., Brostow W., (2012), Combined Effects Of Fly Ash and Waste Ferrochromium on Properties of Concrete, Construction and Building Materials, Vol. 29, pp. 633–640.
12. M. Erdem, H.S. Altundogan, M.D. Turan, and F. Tumen, (2005). Hexavalent Chromium Removal by Ferrochromium Slag, Journal of Hazardous Materials, Vol. 126, pp. 176–182.
13. (a) G. S. Cole and A. M. Sherman, (1995), Lightweight Materials for Automotive Applications, Materials Characterization, Vol. 35, pp. 3–9.

(b) Y Reda, R. Abdel-Karim, and I Elmahallawi. (2008), Improvements in Mechanical and Stress Corrosion Cracking Properties in Al-alloy 7075 via Retrogression and Reaging. Materials Science and Engineering A,;Vol. 485, pp. 468–75.

(c). R.B. Clark, B. Jr Coughran, I. Traina, A. Hernandez, T. Scheck, and C Etuk, (2005), On the correlation of mechanical and physical properties of 7075-T6 Al alloy. Engineering Fail Anal, Vol. 12, pp. 520–6.

(d) S.W. Kim, D.Y. Kim, W.G. Kim, and K. D. Woo. (2001), The Study On Characteristics Of Heat Treatment of the Direct Squeeze Cast 7075 Wrought Al Alloy, Materials Science Engineering A, Vol. 304–306, pp. 721–6.

(e). T.J.A Doel, and P. Bowen, (1996), Tensile Properties of Particulate-Reinforced Metal Matrix Composites, Composites Part A-Appl S, Vol. 27, pp. 655–65.

(f) K. Komai, K. Minoshima, and H. Ryoson, (1993), Tensile and Fatigue Fracture Behavior and Water-Environment Effects in a SiC-Whisker/7075-Aluminum Composite, Composite Science and Technology, Vol. 46, pp. 59–66.
14. A. Macke and B. F. Schultz, (2012) Metal Matrix Composites Offer the Automotive Industry and Opportunity to Reduce Vehicle Weight, Improve Performance, Advanced Materials Processing, Vol. 2, pp. 19–23.
15. S. Kumar, R. Kumar, and A. Bandopadhyay, (2006) Review Innovative Methodologies for the Utilization of Wastes from Metallurgical and Allied Industries, Resources, Conservation and Recycling, Vol. 48, pp. 301-314.
16. C. S. Gahan, M. L. Cunha, and Å. Sandström, (2009), Comparative Study on Different Steel Slags as Neutralizing Agent in Bioleaching, Hydrometallurgy, Vol. 95, pp. 190–197.
17. M. Sánchez, F. Parada, R. Parra, F. Márquez, R. Jara, J. C. Carrasco, and J. Palacios, (2004), Management of Copper Pyrometallurgical Slags: Giving Additional Value to the Copper Mining Industry, in: VII Int. Conference on Molten Slags, Fluxes & Salts, Cape Town, South Africa p. 543–550.
18. N. M. Piatak, M. B. Parsons, and R. R. Seal, (2015), Characteristics and Environmental Aspects of Slag: A Review, Applied Geochemistry, Vol. 57, pp. 236-266.
19. Š. Langová, and D. Matýsek, (2010), Zinc Recovery from Steel-Making Wastes by Acid Pressure Leaching and Hematite Precipitation, Hydrometallurgy, Vol. 101, pp. 171-173.
20. E. A. Vestola, M. K. Kuusenaho, H. M. Närhi, O. H. Tuovinen, J. A. Puhakka, J. J. Plumb, and A. H. Kaksonen, (2010), Acid Bioleaching of Solid Waste Materials from Copper, Steel and Recycling Industries, Hydrometallurgy, Vol. 103, pp. 74-79.
21. A. Baral and R. D. Engelken, (2002), Chromium-based Regulations and Greening in Metal Finishing Industries in the USA, Environmental Science & Policy, Vol. 5, pp. 121–133.
22. K. Bethke, S. Palantöken, V. Andrei, M. Roß, V. S. Raghuwanshi, F. Kettemann, K. Greis, T. T. Ingber, J. B. Stückrath, and S. Valiyaveettil, (2018) Functionalized Cellulose for Water Purification,Antimicrobial Applications, and Sensors, Advanced Functional Materials, Vol. 28, pp. 1800409-14.
23. A. Habib, Haq N. Bhatti and M. Iqbal, (2019), Metallurgical Processing Strategies for Metals Recovery from Industrial Slags, Zeitschrift Fur Physikalische Chemie-International Journal Of Research In Physical Chemistry & Chemical Physics, Vol.xxx, pp.1-31. (https://doi.org/10.1515/zpch-2019-0001).
24. H. Shen, and E. Forssberg, (2003), An Overview of Recovery of Metals from Slags, Waste Management, Vol. 23(10), pp. 933-49 (doi: 10.1016/S0956-053X(02)00164-2).
25. K. Mazurek, (2013), Recovery of Vanadium, Potassium and Iron from A Spent Vanadium Catalyst by Oxalic Acid Solution Leaching, Precipitation and Ion Exchange Processes, Hydrometallurgy Vol. 134-135, pp. 26–31.

26. B. Das, S. Prakash, P. Reddy, and V. Misra, (2007), An overview of Utilization of Slag and Sludge from Steel Industries, Resources, Conservation & Recycling, Vol. 50, pp. 40-57.
27. N. Sano, (2004), Reduction of Chromium Oxide in Stainless Steel Slags, in: Proceedings: TenthInternational Ferroalloys Congress. INFACON X Transformation through Technol., Cape Town p. 670–677.
28. B. Das, J. Mohanty, P. Reddy, and M. Ansari, (1997), Characterizations and Beneficiation Studies of Charge Chrome Slag, Scandinavian Journal of Metallurgy, Vol. 26(4), pp. 153-157.
29. B. R. Choudhury, C. Murty, R. Sripriya, P. Rao, B. Singh, and V. Rao, (1996), Recovery of Metal from Slag/Mixed Metal/Chips Generated in Ferro Alloy Plants, Tata Search (India), Vol. 3, pp. 9-15.
30. L.Holappa, and Y. Xiao, (2004), Slags in Ferroalloys Production – Review of Present Knowledge. Journal of the South African Institute of Mining and Metallurgy, Vol. 104, pp. 429 –438.
31. H.-Y. Li, H.-X. Fang, K. Wang, W. Zhou, Z. Yang, X.-M. Yan, W.-S. Ge, Q.-W. Li, and B. Xie, (2015) Asynchronous Extraction of Vanadium and Chromium from Vanadium Slag by Stepwise Sodium Roasting–Water Leaching, Hydrometallurgy, Vol. 156, pp. 124-135.
32. O. Kononova, G. Bryuzgina, O. Apchitaeva, and Y. Kononov, (2019), Ion Exchange Recovery of Chromium (VI) and Manganese (II) from Aqueous Solutions, Arabian Journal of Chemistry, Vol. 12, pp. 2713-2720 (https://doi.org/10.1016/j.arabjc.2015.05.021).
33. Zhang, X., Gu, F., Peng, Z., Wang, L., Tang, H., Rao, M., Zhang, Y., Li, G., Jiang, T., Wang, Y., 2019. Recovering magnesium from ferronickel slag by vacuum reduction: thermodynamic analysis and experimental verification. ACS Omega 4 (14), 16062–16067.
34. Ostara Nutrient (2022), Recovery, Nutrients | Ostara nutrient recovery [online] Available at: https://ostara.com/nutrients/ [Accessed 8 January 2022].
35. M., Ozdemir, D. Çakır, and I. Kıpçak, (2009), Magnesium Recovery from Magnesite Tailings by Acid Leaching and Production of Magnesium Chloride Hexahydrate from Leaching Solution by Evaporation. International Journal of Mineral Processing, Vol. 93, pp. 209–212.
36. N. Raza, Z. Zafar, and N.-U. Haq, (2013), An Analytical Model Approach for the Dissolution Kinetics of Magnesite Ore using Ascorbic Acid as Leaching Agent, International Journal of Metal, Vol. xxx, pp. 1–6.
37. Z. Bradford-Hartke, A. Razmjou, and L. Gregory, (2021), Factors Affecting Phosphorus Recovery as Struvite: Effects of Alternative Magnesium Sources, Desalination, Vol. 504, pp. 114949-xxx.
38. S. Shaddel, T. Grini, J. Andreassen, S. Østerhus, and S. Ucar, (2020), Crystallization Kinetics and Growth of Struvite Crystals by Seawater Versus Magnesium Chloride as Magnesium Source: Towards Enhancing Sustainability and Economics of Struvite Crystallization, Chemosphere Vol. 256, pp. 126968-xxx.

39. R. M. Reynolds and M. Olper, (1990), The Engitec System for Treatment of Salt Slag from Secondary Aluminum Smelters, The Minerals, Metals & Materials Society, Warrendale, PA, Vol. xxx, pp. 439-xxx.
40. The Use of Steel Slag in Passive Treatment Design for Amd Discharge in the Huff Run Watershed Restoration, James Hamilton, Jim Gue, and Cheryl Socotch 2, Paper was presented at the 2007 National Meeting of the American Society of Mining and Reclamation, Gillette, WY, 30 Years of SMCRA and Beyond June 2-7, 2007. R.I. Barnhisel (Ed.) Published by ASMR, 3134 Montavesta Rd., Lexington, KY 40502, Vol. xxx, pp. 272-282.
41. N. M. Piatak, R. R. Seal, D. A. Hoppe, C. J. Green, and P. M. Buszka, (2019), Geochemical Characterization of Iron and Steel Slag and Its Potential to Remove Phosphate and Neutralize Acid, Minerals, Vol. 9, pp. 468-26 (https://doi.org/10.3390/min9080468).
42. https://www.dreamstime.com/photos-images/rip-rap.html
43. https://www.istockphoto.com/photos/riprap
44. M. J. Rezaei-Hosseinabadi, M. Bayat, B. Nadi, and A. Rahimi, (2022), Case Study Sustainable Utilization of Steel Slag as Granular Column for Ground Improvement in Geotechnical Projects, Case Studies in Construction Materials, Vol. 17, pp. e01333.
45. https://www.harsco-environmental.com/resources/white-papers/use-steel-slag-aggregate-stone-column-ground-stabilization
46. https://www.istockphoto.com/photo/slag-from-burnt-coal-gm1219565715-356783309
47. https://www.alamy.com/stock-photo/slag-stones.html?sortBy=relevant

Index

Page numbers followed by f and t indicate figures and tables, respectively.

Also of Interest

Check out the previously published books by the same authors

Development of Geopolymer from Pond Ash-Thermal Power Plant Waste

Novel Constructional Materials for Civil Engineers

The book highlights the mechanism of the formation of GP from pond ash. Properties of structures made with GP concrete are found to be comparable to those made with cement concrete. Systematic investigations are presented to understand the chemistry of GP formation with pond ash materials. Performances of these materials above ambient temperature, as well as with different environmental conditions, are also evaluated.

Published 2023. ISBN 978-1-394-16652-7

High Electrical Resistance Ceramics

Thermal Power Plant Waste Resources

This book highlights the preparation of ceramics using pond/fly ash. Since the mullite phase formed by heat treatment improves the properties of ceramics, current investigations will perhaps be the first attempt to develop ceramics using pond ash. The properties of components made with these developed ceramics are found to be comparable to those made with porcelain. The extensively reviewed chapters of this book illustrate the current status of research on these materials. At the end of each of the 10 chapters, conclusions are drawn which will benefit researchers working in this area. Subjects discussed include:

- The fundamentals of thermal power plant wastes;
- Different production methods of ceramics and various characterization techniques;
- The preparation of ceramics from fly ash and fly ash/kaolin composite;

- The production of ceramics using pond ash;
- The preparation and characterization of geopolymer from pond ash and the preparation of pond ash composite;
- Production of ceramic matrix composite (CMC) using pond ash and pyrophyllite;
- The preparation of ceramics using pond ash and k-feldspar mixture.

Published 2023. ISBN 978-1-394-19993-8

www.scrivenerpublishing.com